H. C. (Hans Christian) Andersen

The Sand-Hills of Jutland

H. C. (Hans Christian) Andersen

The Sand-Hills of Jutland

ISBN/EAN: 9783744709422

Printed in Europe, USA, Canada, Australia, Japan

Cover: Foto ©berggeist007 / pixelio.de

More available books at **www.hansebooks.com**

THE

SAND-HILLS OF JUTLAND.

BY

HANS CHRISTIAN ANDERSEN,

AUTHOR OF THE "IMPROVISATORE," ETC.

BOSTON:

TICKNOR AND FIELDS.

M DCCC LX.

University Press, Cambridge:
Electrotyped and Printed by Welch, Bigelow, & Co.

CONTENTS.

The Sand-hills of Jutland.

THIS is a story from the Jutland sand-hills, but it does not commence there; on the contrary, it commences far away towards the south, in Spain. The sea is the highway between the two countries. Fancy yourself there. The scenery is beautiful; the climate is warm. There blooms the scarlet pomegranate amidst the dark laurel trees; from the hills a refreshing breeze is wafted over the orange groves and the magnificent Moorish halls, with their gilded cupolas and their painted walls. Processions of children parade the streets with lights and waving banners; and, above these, clear and lofty rises the vault of heaven, studded with glittering stars. Songs and castanets are heard; youths and girls mingle in the dance under the blossoming acacias; whilst beggars sit upon the sculptured blocks of marble, and refresh themselves with the juicy water-melon. Life dozes here: it is all like a charming dream, and one indulges in it. Yes, thus did two young newly-married persons, who also possessed all the best gifts of earth — health, good humour, riches, and rank.

"Nothing could possibly exceed our happiness," they

1   A

said in the fulness of their joyful hearts; yet there was
one degree of still higher happiness to which they might
attain, and that would be when God blessed them with a
child — a son, to resemble them in features and in disposition.

That fortunate child would be hailed with rapture;
would be loved and daintily cared for; would be the heir
to all the advantages that wealth and high birth can
bestow.

The days flew by as a continual festival to them.

"Life is a merciful gift of love — almost inconceivably
great," said the young wife; "but the fulness of this
happiness shall be tasted in that future life, when it will
increase and exist to all eternity. The idea is incomprehensible to me."

"That is only an assumption among mankind," said
her husband. "In reality, it is frightful pride and over-
weening arrogance to think that we shall live for ever —
become like God. These were the serpent's wily words,
and he is the father of lies."

"You do not, however, doubt that there is a life after
this one?" asked his wife; and for the first time a cloud
seemed to pass over their sunny heaven of thought.

"Faith holds forth the promise of it, and the priests
proclaim it," said the young man; "but, in the midst of
all my happiness, I feel that it would be too craving, too
presumptuous, to demand another life after this one — a
happiness to be continual. Is there not so much granted
in this existence that we might and ought to be content
with it?"

"To us — yes, there has been much granted," replied

the young wife; "but to how many thousands does not this life become merely a heavy trial? How many are not, as it were, cast into this world to be the victims of poverty, wrangling, sickness, and misfortune? Nay, if there were no life after this one, then everything in this globe has been unequally dealt out; then God would not be just."

"The beggar down yonder has joys as great, to his ideas, as are those of the monarch in his splendid palace to him," said the young man; "and do you not think that the beasts of burden, which are beaten, starved, and toiled to death, feel the oppressiveness of their lot? They also might desire another life, and call it unjust that they had not been placed amidst a higher grade of beings."

"In the kingdom of heaven there are many mansions, Christ has told us," answered the lady. "The kingdom of heaven is infinite, as is the love of God. The beasts of the field are also His creation; and my belief is that no life will be extinguished, but will win that degree of happiness which may be suitable to it, and that will be sufficient."

"Well, this world is enough for me," said her husband, as he threw his arms round his beautiful, amiable wife, and smoked his cigarette upon the open balcony, where the deliciously cool air was laden with the perfume of orange trees and beds of carnations. Music and the sound of castanets arose from the street beneath; the stars shone brightly above; and two eyes full of affection, the eyes of his charming wife, looked at him with love which would live in eternity.

"Such moments as these," he exclaimed, "are they

not well worth being born for — born to enjoy them, and then to vanish into nothingness?"

He smiled; his wife lifted her hand and shook it at him with a gesture of mild reproach, and the cloud had passed over — they were too happy.

Everything seemed to unite for their advancement in honour, in happiness, and in prosperity. There came a change, but in place — not in anything to affect their well-being, to damp their joy, or to ruffle the smooth current of their lives. The young nobleman was appointed by his king ambassador to the court of Russia. It was a post of honour to which he was entitled by his birth and education. He had a large private fortune, and his young wife had brought him one not inferior to his own, for she was the daughter of one of the richest men in the kingdom. A large ship was about that time to go to Stockholm. It was selected to convey the rich man's dear daughter and son-in-law to St. Petersburg; and its cabin was fitted up as if for the use of royalty — soft carpets under the feet, silken hangings, and every luxury around.

Amidst the ancient Scandinavian ballads, known to all Danes under their general title of *Kæmpeviser*, there is one called "The King of England's Son.' He likewise sailed in a costly ship; its anchor was inlaid with pure gold, and every rope was of twisted silk. Every one who saw the Spanish vessel must have remembered the ship in this legend, for there was the same pageantry, the same thoughts on their departure.

"God, let us meet again in joy!"

The wind blew freshly from off the Spanish shore, and

the last adieux were therefore hurried; but in a few weeks they would reach their destination. They had not gone far, however, before the wind lulled, the sea became calm, its surface sparkled, the stars above shone brightly, and all was serenity in the splendid cabin.

At length they became tired of the continued calm, and wished that the breeze would rise and swell into a good strong wind, if it would only be fair for them; but they still lacked wind, and if it did arise, it was always a contrary one. Thus passed weeks, and when at length the wind became fair, and blew from the south-west, they were half way between Scotland and Jutland. Just then the wind shifted, and increased to a gale, as it is described to have done in the ballad of " The King of England's Son."

> " The sky grew dark, and the wind it blew,
>   They could see neither land nor haven of rest;
> So then they cast out their anchor true,
>   But to Denmark they drove with the gale from the west."

This was many years ago. King Christian the Seventh occupied the Danish throne, and was then a young man. Much has happened since that time, much has changed; lakes and morasses have become fruitful meadows, wild moors have become cultivated land, and on the lee of the West Jutlander's house grow apple trees and roses; but they must be sheltered from the sharp west winds. Up there one can still, however, fancy one's self back in the period of Christian the Seventh's reign. As then in Jutland, so even now, stretch for miles and miles the brown heaths, with their tumuli, their meteors, their knolly, sandy cross roads. Towards the west,

where large streams fall into the fiords, are to be seen
wide plains and bogs, encircled by high hills, which, like
a row of Alpine mountains with pinnacles formed like
saws, frown over the sea, which is separated from them
only by high clay banks; and year after year the sea
bites a large mouthful off of these, so that their edges
and summits topple over as if shaken by an earthquake.
Thus they look at this day, and thus they were many
years ago, when the happy young couple sailed from
Spain in the magnificent ship.

It was the end of September. It was Sunday and
sunshine: the sound of the church bells reached afar,
even to Nissumfiord. The churches up there were like
rocks with spaces hewn out in them: each one of them
was like a piece of a mountain, so heavy and massive.
The German Ocean might have rolled over them, and
they would have stood firmly. Many of them had no
spires or towers, and the bells hung out in the open air
between two beams. The church service was over.
The congregation had passed from the house of God out
into the churchyard, where then, as now, not a tree, not
a bush was to be seen — not a single flower, not a gar-
land laid upon a grave. Little knolls or heaps of earth
point out where the dead are buried; a sharp kind of
grass, lashed by the wind, grows over the whole churchyard. A solitary grave here and there has, perhaps, a
monument; that is to say, the mouldering trunk of a
tree, rudely carved into the shape of a coffin. The
pieces of tree are brought from the woods of the west.
The wild ocean provides, for the dwellers on the coast,
beams, planks, and trees, which the dashing billows cast

upon the shore. The wind and the sea spray soon decay
these tree monuments. Such a stump was lying over
the grave of a child, and one of the women who had
come out of the church went towards it. She stood
gazing upon the partially loosened piece of wood.
Shortly afterwards her husband joined her. They remained for a time without either of them uttering a
single word; then he took her hand, and led her from the
grave out upon the heath, across the moor, in the direction of the sand-hills. For a long time they walked
in silence. At last the husband said, —

"It was an excellent sermon to-day. If we had not
our Lord we should have nothing."

"Yes," said the wife, "He sends joy, and He sends
affliction. He is right in all things. To-morrow our little boy would have been five years old if he had been
spared to us."

"There is no use in your grieving for his loss," replied
the husband. "He has escaped much evil. He is now
where we must pray to be also received."

They dropped the painful subject, and pursued their
way towards their house amidst the sand-hills. Suddenly, from one of these where there was no lyme-grass
to keep down the sand, there arose as it were a thick
smoke. It was a furious gust of wind, that had pierced
the sand-hill, and whirled about in the air the fine parti-
cles of sand. The wind veered round for a minute; and
all the dried fish that was hung up on cords outside of the
house knocked against its walls, then everything was still
again. The sun was shining warmly.

The man and his wife entered their house, and having

soon divested themselves of their Sunday clothes, they
hastened over the sand-hills, which stood like enormous
waves of sand suddenly arrested in their course. The
sea-reed's and the lyme-grass's blue-green sharp blades
gave some variety to the white sand. Some neighbours
joined the couple who had just come from church, and
they assisted each other in dragging the boats higher up
the beach. The gale was increasing; it was bitterly
cold; and when they were returning over the hills, the
sand and small stones whisked into their faces, the waves
mounted high with their white crests, and the spray
dashed after them.

It was evening; there was a doleful whistling in the
air, increasing every moment — a wild howling, as if a
host of unseen despairing spirits were uttering their complaints. The moaning sound overpowered even the
angry dashing of the waves, although the fisherman's
house lay so near to the shore. The sand drifted against
the windows, and every now and then came a blast that
shook the house to its foundation. It was very dark, but
the moon would rise at midnight.

The air cleared; yet the storm still raged in all its
might over the deep gloomy sea. The fishermen and
their families had retired for some time to rest, but no
one could close his eyes in such terrible weather. Some
one knocked at the windows of some of the cottages, and
when the doors were opened the person said, —

"A large ship is lying fast upon the outer shoal."

In a moment the fishermen and their wives were up
and dressed.

The moon had risen, and there was light enough to see

if they had not been blinded by the sand that was flying about. The wind was so strong that they were obliged to lie down, and creep amidst the gusts over the sandhills; and there flew through the air, like swan's down, the salt foam and spray from the sea, which, like a roaring, boiling cataract, dashed upon the beach. A practised eye was required to discern quickly the vessel outside. It was a large ship; it was lifted a few cable lengths forward, then driven on towards the land, struck upon the inner sand-bank, and stood fast. It was impossible to go to the assistance of the ship, the sea was running too high: it beat against the unfortunate vessel, and dashed over her. The people on shore thought that they heard cries of distress — cries of those in the agony of death; and they saw the desperate, useless activity on board. Then came a sea that, like a crushing avalanche, fell upon the bowsprit, and it was gone. The stern of the vessel rose high above the water — two people sprang from it together into the sea — a moment, and one of the most gigantic billows that were rolling up against the sand-hills cast a body upon the shore: it was that of a female, and every one believed it was a corpse. Two women, however, knelt down by the body, and thinking that they found in it some sign of life, it was carried over the sand-hills to a fisherman's house. How beautiful she was, and how handsomely dressed! — evidently a lady of rank.

They placed her in the humble bed; there was no linen on it, only blankets to wrap her in, yet these were very warm.

She soon came to life, but was in a high fever. She

1*

did not seem to know what had happened, or to remark
where she was; and this was probably fortunate, since all
who were dear to her on board the ill-fated ship were
lying at the bottom of the sea. It had been with them
as described in the song, " The King of England's
Son:" —

> " It was, in sooth, a piteous sight!
> The ship broke up to bits that night."

Portions of the wreck were washed ashore. She was
the only living creature out of all that had so lately
breathed and moved on board the doomed ship. The
wind was howling their requiem over the inhospitable
coast. For a few minutes she slept peacefully, but soon
she awoke and uttered groans of pain; she cast up her
beautiful eyes towards heaven, and said a few words,
but no one there could understand them.

Another helpless being soon made its appearance, and
her new-born babe was placed in her arms. It ought to
have reposed on a stately couch, with silken curtains, in
a splendid house. It ought to have been welcomed with
joy to a life rich in all this world's goods; but our Lord
had ordained that it should be born in a peasant's hut, in
a miserable nook. Not even one kiss did it receive from
its mother.

The fisherman's wife laid the infant on its mother's
breast, and it rested near her heart; but that heart had
ceased to beat — she was dead! The child who should
have been nurtured amidst happiness and wealth was
cast a stranger into the world — thrown up by the sea
among the sand-hills, to experience heavy days and the
fate of the poor. And again we call to mind the old
song : —

" The king's son's eyes with big tears fill:
' Alas! that I came to this robber-hill.
Here nothing awaits me but evil and pain.
Had I haply but come to Herr Buggé's domain,
Neither knight nor squire would have treated me ill.' "

A little to the south of Nissumfiord, on that portion of the shore which Herr Buggé had formerly called his, the vessel had stranded. Those rough, inhuman times, when the inhabitants of the west coast dealt cruelly, it is said, with the shipwrecked, had long passed away; and now the utmost compassion was felt, and the kindest attention paid to those whom the engulfing sea had spared. The dying mother and the forlorn child would have met with every care wherever " the wild wind had blown ;". but nowhere could they have been received with more cordial kindness than by the poor fishwife who, only the previous morning, had stood with a heavy heart by the grave wherein reposed her child, who on that very day would have attained his fifth year if the Almighty had permitted him to live.

No one knew who the foreign dead woman was, or whence she came. The broken planks and fragments of the ship told nothing.

In Spain, at that opulent house, there never arrived either letter or message from the daughter and son-in-law; they had not reached their destination; fearful storms had raged for some weeks. They waited with anxiety for months. At last they heard, "Totally lost — every one on board perished ! "

But at Huusby-Klitter, in the fisherman's cottage, there dwelt now a little urchin.

Where God bestows food for two, there is always

something for a third; and near the sea there is plenty
of fish to be found.   The little stranger was named
Jörgen.

"He is surely a Jewish child," said some people, "he
has so dark a complexion."

"He may, however, be an Italian or a Spaniard," said
the priest.

The whole tribe of fishermen and women comforted
themselves that, whatever was his origin, the child had
received Christian baptism.   The boy throve, his noble
blood mantled in his cheek, and he grew strong, notwithstanding poor living.   The Danish language, as it is
spoken in West Jutland, became his mother tongue.
The pomegranate seed from the Spanish soil became the
coarse grass on the west coast of Jutland.   Such are the
vicissitudes of life!

To that home he attached himself with his young life's
roots.   Hunger and cold, the poor man's toil and want,
he was to experience, but also the poor man's joys.

Childhood has its bright periods, which shine in recollection through the whole of after life.   How much had
he not to amuse him, and to play with!   The entire
seashore, for miles in length, was covered with playthings
for him — a mosaic of pebbles red as coral, yellow as
amber, and pure white, round as birds' eggs, all smoothed
and polished by the sea.   Even the scales of the dried
fish, the aquatic plants dried by the wind, the shining
seaweed fluttering among the rocks — all were pleasant
to his eye, and matter for his thoughts; and the boy was
an excitable, clever child.   Much genius and great abili-
ities lay dormant in him.   How well he remembered all

the stories and old ballads he heard; and he was very quick with his fingers. With stones and shells he would plan out whole scenes he had heard as if in a picture: one might have ornamented a room with these handiworks of his. "He could cut out his thoughts with a stick," said his foster-mother; and yet he was but a little boy. His voice was very sweet — melody seemed to have been born with him. There were many finely-toned strings in that breast; they might have sounded forth in the world had his lot been otherwise cast than in a fisherman's house on the shores of the German Ocean.

One day a ship foundered near. A case was thrown up on the land containing a number of flower-bulbs. Some took them and put them into their cooking pots, thinking they were to be eaten; others were left to rot upon the sand; none of them fulfilled their destination — to unfold the lovely colours, the beauty that lay in them. Would it be better with Jörgen? The poor flower-roots were soon done for; there might be years of trial before him.

It never occurred to him, or to any of the people around him, to think their days lonely and monotonous: there was abundance to do, to hear, and to see. The ocean itself was a great book; every day he read a new page in it — the calm, the swell of the sea, the breeze, the storm. The beach was his favourite resort; going to church was his event, his visit of importance, though of visits there was one which occasionally took place at the fisherman's house that was particularly welcome to him. Twice a year his foster-mother's brother, the eel-man from Fjaltring, up near Rovbierg, paid them a visit. He came in a painted cart full of eels. The cart was closed and locked like a chest, and painted with blue, red, and white tulips; it was drawn by two dun-coloured bullocks, and Jörgen was allowed to drive them.

The eel-man was a very good-natured, lively guest. He always brought a keg of brandy with him; every one got a dram of it, or a coffee-cup full if glasses were scarce; even Jörgen, though he was but a little fellow, was treated to a good thimbleful. That was to keep down the fat eels, said the eel-man; and then he never failed to tell a story he had often told before, and, when people laughed at it, he immediately told it over again to the same persons; but this is a habit with all talkative individuals; and as Jörgen, during the whole time that he. was growing up, and into the years of his manhood, often quoted phrases in this story, and applied them to himself, we may as well listen to it.

" Out in the rivulet dwelt eels, and the eel-mother said to her daughters, when they begged to be allowed to go a little way alone up the stream. ' Do not go far, lest the horrible eel-spearer should come, and take you all away.

" But they went very far, and of eight daughters only three returned to their mother, and these came wailing, ' We only went a short way from the door, when the terrible eel-spearer came and killed our five sisters.' ' They will come back again,' said the eel-mother. ' No,' said the daughters, ' for he skinned them, cut them in pieces, and fried them.' ' They will come again,' repeated the mother. ' Impossible, for he ate them.' ' They will

come again,' still persisted the eel-mother. But he
drank brandy after he had eaten them,' said the daughter. 'Did he? Oh! oh! then they will never come
again,' howled the mother. 'Brandy buries eels.'

"And therefore one must always drink a little brandy
after that dish," said the eel-man.

And this story made a great impression on little Jörgen, and partly influenced his life. He took the tinsel
for the gold. He also wished to go "a little way up the
stream"—that is to say, to go away in a ship to see the
world—and his mother said as the eel-mother had done,
"There are many bad men—eel-spearers." But a little
way beyond the sand-hills, and a little way on the heath,
he was allowed to go, he begged so hard. Four happy
days, however—days that seemed the brightest among
his childish years, turned up: he was to go to a large
meeting. What pleasure, although it was to a funeral!

A relation of the fisherman's family, who had been in
easy circumstances, was dead. The farm lay inland—
"eastward, a little to the north," it was said. The father
and mother were both going, and Jörgen was to accompany them. On leaving the sand-hills, they passed over
heaths and boggy lands, until they came to the green
meadows where Skjærumaa winds its way—the river
with the numerous eels, where the eel-mother with her
daughters lived, those whom the cruel man speared and
cut in pieces, though there were men who had scarcely
treated their fellow-men better. Even Herr Buggé, the
knight who was celebrated in the old song, was murdered
by a wicked man; and though he was himself called so
good, he wished to put to death the builder who had built

for him his castle, with its tower and thick walls, just
where Jörgen and his foster-parents stood, where Skjærumaa falls into the Nissumfiord. The sloping bank or
ascent to the ramparts was still to be seen, and red fragments of the walls still marked out the circumference of
the ancient building. Here had Herr Buggé, when the
builder had taken his departure, said to his squire,
"Follow him, and say, Master, the tower leans to one
side If he turns, slay him on the spot, and take the
money from him that he got from me; but, if he does not
turn, let him go on in peace." And the squire overtook
the builder, and said what he was ordered to say ; and
the builder replied, "The tower does not lean to one side,
but by and by there will come from the westward one in
a blue cloak, and *he* will make it bend." A hundred
years afterwards this prediction was fulfilled, for the German Ocean rushed in, and the tower fell; but the then
owner of the property, Prebjörn Gyldenstierne, erected
a habitation higher up, and that stands now, and is called
Nörre-Vosborg.

Jörgen, with his foster-parents, had to pass this place.
Of every little town hereabout he had heard stories during the long winter evenings; now he saw the castle,
with its double moats, its trees and bushes, its ramparts
overgrown with bracken. But the most beautiful sight
was the lofty linden trees, that filled the air with so sweet
a perfume. Towards the north-west, in a corner of the
garden, stood a large bush with flowers that were like
winter's snow amidst summer's green. It was an elder
tree, the first Jörgen had ever seen in bloom. That and
the linden trees were always remembered during his

future years as Denmark's sweetest perfume and beauty, which the soul of childhood " for the old man laid by."

The journey soon became more extended, and the country less wild. After passing Nörre-Vosborg, where the elder tree was in bloom, he had the pleasure of travelling in a sort of carriage, for they met some of the other guests who were going to the funeral feast, as it might be called, and were invited into their conveyance. To be sure they had all three to stuff themselves into a very narrow back seat, but that was better, they thought, than walking. They drove over the uneven heaths; the bullocks which drew their cart stopped whenever they came to a little patch of green grass among the heather. The sun was shining warmly, and it was wonderful to see, far in the distance, a smoke that undulated, yet was clearer than the air — one could see through it: it was as if rays of light were rolling and dancing over the heath.

"It is the Lokéman, who is driving his sheep," was told Jörgen, and that was enough for him. He fancied he was driving into the land of marvellous adventures and fairy tales; yét he was only amidst realities. How still it was there!

Far before them stretched the heath, but it looked like a beautifully variegated carpet; the ling was in flower, the Cyprus-green juniper bushes and the fresh oak shoots seemed like bouquets among the heather. But for the many poisonous vipers, how delightful it would have been to roll about there! The party spoke of them, and of the numerous wolves that had abounded in that neighbourhood, on account of which the district was called Ulvborg-Herred. The old man who was driving related how, in his father's time, the horses had often to fight a hard battle with these now extirpated wild animals; and that one morning, on coming out, he found one of his horses treading upon a wolf he had killed; but the flesh was entirely stripped from the horse's legs.

Too quickly for Jörgen did they drive over the uneven heath, and through the deep sand. They stopped at length before the house of mourning, which was crowded with strangers, some inside, some on the outside. Vehicle after vehicle stood together; the horses and oxen were turned out amidst the meagre grass; large sandhills, like those at home by the German Ocean, were to be seen behind the farm, and stretched far away in wide long ranges. How had they come there, twelve miles inland, and nearly as high and as large as those near the shore? The wind had lifted them and removed them: they also had their history.

Psalms were sung, and tears were shed by some of the old people, otherwise all was very pleasant, thought Jörgen. Here was plenty to eat and drink — the nicest fat eels; and it was necessary to drink brandy-snaps after eating them, "to keep them down," the eel-man had said; and his words were acted upon here with all due honour.

Jörgen was in, and Jörgen was out. By the third day he felt himself as much at home here as he had done in the fisherman's cottage, where he had lived all his earlier days. Up here on the heath it was different from down there, but it was very nice. It was covered with heather-bells and bilberries; they were so large and so sweet;

one could mash them with one's foot, so that the heather should be dripping with the red juice. Here lay one tumulus, there another; columns of smoke arose in the calm air; it was the heath on fire, they said, it shone brightly in the evening.

The fourth day came, and the funeral solemnities were over — the fisherman and his family were to leave the land sand-hills for the strand sand-hills.

"Ours are the largest though;" said the father, "these are not at all important-looking."

And the conversation fell on how they came there, and it was all very intelligible and very rational. A body had been found on the beach, and the peasants had buried it in the churchyard; then commenced a drifting of sand — the sea broke wildly on the shore, and a man in the parish who was noted for his sagacity advised that the grave should be opened, to ascertain if the buried corpse lay and sucked his thumb; for if he did that, it was a merman whom they had buried, and the sea would force its way up to take him back. The grave was accordingly opened, and lo! he they had buried was found sucking his thumb; so they took him up instantly, placed him on a car, harnessed two oxen to it, and dragged him over heaths and bogs out to the sea; then the sand drift stopped, but the sand-hills have always remained. To all this Jörgen listened eagerly; and he treasured this ancient legend in his memory, along with all that had happened during the pleasantest days of his childhood — the days of the funeral feast.

It was delightful to go from home, and to see new places and new people; and he was to go still farther

away. He went on board a ship. He went forth to see
what the world produced; and he found bad weather,
rough seas, evil dispositions, and harsh masters. He
went as a cabin-boy! Poor living, cold nights, the rope's
end, and hard thumps with the fist were his portion.
There was something in his noble Spanish blood which
always boiled up, so that angry words rose often to his
lips; but he was wise enough to keep them back, and he
felt pretty much like an eel being skinned, cut up, and
laid on the pan.

" I will come again," said he to himself. The Spanish
coast, his parents' native land, the very town where they
had lived in grandeur and happiness, he saw; but he
knew nothing of kindred and a paternal home, and his
family knew as little of him.

The dirty ship-boy was not allowed to land for a long
time, but the last day the ship lay there he was sent on
shore to bring off some purchases that had been made.

There stood Jörgen in wretched clothes, that looked
as if they had been washed in a ditch and dried in the
chimney: it was the first time that he, a denizen of the
solitary sand-hills, had seen a large town. How high
the houses were, how narrow the streets, swarming with
human beings; some hurrying this way, others going
that way — it was like a whirlpool of townspeople, peasants, monks, and soldiers. There were a rushing along,
a screaming, a jingling of the bells on the asses and the
mules, and the church bells ringing too. There were to
be heard singing and babbling, hammering and banging;
for every trade had its workshop either in the doorway
or on the pavement. The sun was burning hot, the air

was heavy : it was as if one had entered a baker's oven full of beetles, lady-birds, bees, and flies, that hummed and buzzed. Jörgen scarcely knew, as the saying is, whether he was on his head or his heels.

Then he beheld, at a little distance, the immense portals of the cathedral; light streamed forth from the arches that were so dim and gloomy above; and there came a strong scent from the incense. Even the poorest, most tattered beggars ascended the wide stairs to the church, and the sailor who was with Jörgen showed him the way in. Jörgen stood in a sacred place; splendidly-painted pictures hung round in richly-gilded frames; the holy Virgin, with the infant Jesus in her arms, was on the altar amidst flowers and light; priests in their magnificent robes were chanting; and beautiful, handsomely-dressed choristers swung backwards and forwards silver censers. There was in everything a splendour, a charm, that penetrated to Jörgen's very soul, and overwhelmed him. The church and the faith of his parents and his ancestors surrounded him, and touched a chord in his heart which caused tears to start to his eyes.

From the church they proceeded to the market. He had many articles of food and matters for the use of the cook, to carry. The way was long, and he became very tired; so he stopped to rest outside of a large handsome house, that had marble pillars, statues, and wide stairs. He was leaning with his burden against the wall, when a finely-bedizened porter came forward, raised his silver-mounted stick to him, and drove him away — him, the grandchild of its owner, the heir of the family; but none there knew this, nor did he himself.

He returned on board, was thumped and scolded, had little sleep and much work. Such was his life! And it is very good for youth to put up with hard usage, it is said. Yes, if it makes age good.

The period for which he had been engaged was expired — the vessel lay again at Ringkiöbingfiord. He landed, and went home to Huusby-Klitter; but his mother had died during his absence.

The winter which followed was a severe one. Snow storms drove over sea and land: one could scarcely face them. How differently were not things dealt out in this world! Such freezing cold and drifting snow here, whilst in Spain was burning heat, almost too great; and yet when, one clear, frosty day at home, Jörgen saw swans flying in large flocks from the sea over Nissumfiord, and towards Nörre-Vosborg, he thought that the course they pursued was the best, and all summer pleasures were to be found there. In fancy he saw the heath in bloom, and mingling with it the ripe, juicy berries; the linden trees and elder bushes at Nörre-Vosborg were in flower. He must return there yet.

Spring was approaching, the fishing was commencing, and Jörgen lent his help. He had grown much during the last year, and was extremely active. There was plenty of life in him; he could swim, tread the water, and turn and roll about in it. He was much inclined to offer himself for the mackerel shoals: they take the best swimmer, draw him under the water, eat him up, and so there is an end of him; but this was not Jörgen's fate.

Among the neighbours in the sand-hills was a boy named Morten. He and Jörgen left the fishing, and they both hired themselves on board a vessel bound to Norway, and went afterwards to Holland. They were always at odds with each other, but that might easily happen when people were rather warm-tempered; and they could not help showing their feelings sometimes in expressive gestures. This was what Jörgen did once on board when they came up from below quarrelling about something. They were sitting together, eating out of an earthen dish they had between them, when Jörgen, who was holding his clasp-knife in his hand, raised it against Morten, looking at the moment as white as chalk, and ghastly about the eyes. Morten only said, —

"So you are of that sort that will use the knife!"

Scarcely had he uttered these words before Jörgen's hand was down again; he did not say a syllable, ate his dinner, and went to his work; but when he had finished that, he sought Morten, and said, —

"Strike me on the face if you will — I have deserved it. There is something in me that always boils up so."

"Let bygones be bygones," said Morten; and thereupon they became much better friends. When they returned to Jutland and the sand-hills, and told all that had passed, it was remarked that Jörgen might boil over, but he was an honest pot for all that.

"But not of Jutland manufacture — he cannot be called a Jutlander," was Morten's witty reply.

They were both young and healthy, well-grown, and strongly built, but Jörgen was the most active.

Up in Norway the country people repair to the summer pastures among the mountains, and take their cattle there to grass. On the west coast of Jutland, among the sand-hills, are huts built of pieces of wrecks, and covered with peat and layers of heather. The sleeping-places stretch round the principal room; and there sleep and live, during the early spring time, the people employed in the fishing. Every one has his *Æsepige*, as she is called, whose business it is to put bait on the hooks, to await the fishermen at their landing-place with warm ale, and have their food ready for them when they return weary to the house. These girls carry the fish from the boats, and cut them up; in short, they have a great deal to do.

Jörgen, his father, and a couple of other fishermen, with their *Æsepiger*, or serving girls, were together in one house. Morten lived in the house next to theirs.

There was one of these girls called Elsé, whom Jörgen had known from her infancy. They were great friends, and much alike in disposition, though very different in appearance. He was of a dark complexion, and she was very fair, with hair almost of a golden colour; her eyes were as blue as the sea when the sun is shining upon it.

One day when they were walking together, and Jörgen was holding her hand with a tight and affectionate grasp, she said to him, —

"Jörgen, I have something on my mind. Let me be be your *Æsepige*, for you are to me like a brother; but Morten, who has hired me at present — he and I are sweethearts. Do not mention this, however, to any one."

And Jörgen felt as if a sand-hill had opened under

him. He did not utter a single word, but nodded his head by way of a yes — more was not necessary; but he felt suddenly in his heart that he could not endure Morten, and the longer he reflected on the matter the clearer it became to him. Morten had stolen from him the only one he cared for, and that was Elsé. She was now lost to him.

If the sea should be boisterous when the fishermen return with their little smacks, it is curious to see them cross the reefs. One of the fishermen stands erect in advance, the others watch him intently, while sitting with their oars ready to use when he gives them a sign that now are coming the great waves which will lift the boats over; and they are lifted, so that those on shore can only see their keels. The next moment the entire boat is hidden by the surging waves — neither boat, nor mast, nor people are to be seen: one would fancy the sea had swallowed them up. A minute or two more, and they show themselves, looking as if some mighty marine monsters were creeping out of the foaming sea, the oars moving like their legs. With the second and the third reef the same process takes place as with the first; and now the fishermen spring into the water and drag the boats on shore, every succeeding billow helping and giving them a good lift until they are fairly out of the water. One false move on the outside of the reefs — one moment's delay, and they would be shipwrecked.

"Then it would be all over with me, and with Morten at the same time." This thought came across Jörgen's mind out at sea, where his foster-father had been taken

2

suddenly ill: he was in a high fever. This was just a little way from the outer reef. Jörgen sprang up.

"Father, allow me," he cried, and his eye glanced over Morten and over the waves; but just then every oar was raised for the great struggle, and as the first enormous billow came, he observed his father's pale suffering countenance, and he could not carry out the wicked design that had suggested itself to his mind. The boat got safely over the reefs, and in to the land; but Jörgen's evil thoughts remained, and his blood boiled at every little disagreeable act that started up in his recollection from the time that he and Morten had been comrades, and his anger increased as he remembered each offence. Morten had supplanted him, he felt assured of that; and that was enough to make him hateful to him. A few of the fishermen remarked his scowling looks at Morten, but Morten himself did not; he was just as usual, ready to give every assistance, and very talkative — a little too much of the latter, perhaps.

Jörgen's foster-father was obliged to keep his bed; he became worse, and died within a week; and Jörgen inherited the house behind the sand-hills — a humble habitation to be sure, but it was always something. Morten had not so much.

"You will not take service any more, Jörgen, I suppose, but will remain among us now," said one of the old fishermen.

But Jörgen had no such intention. He was thinking, on the contrary, of going away to see a little of the world. The eel-man of Fjaltring had an uncle up at Gammel-Skagen; he was a fisherman, but also a thriving trader who owned some little vessels. He was such an excellent old man, it would be a good thing to take service with him. Gammel-Skagen lies on the northern part of Jutland, at the other extremity of the country from Huusby-Klitter, and that was what Jörgen thought most of. He was determined not to stay for Elsé and Morten's wedding, which was to take place in a couple of weeks.

"It was foolish to take his departure now," was the opinion of the old fisherman who had spoken to him before. "Now Jörgen had a house, Elsé would most likely prefer taking him."

Jörgen answered so shortly, when thus spoken to, that it was difficult to ascertain what he thought; but the old man brought Elsé to him. She did not say much; but this she did say, —

"You have now a house: one must take that into consideration.

And Jörgen also took much into consideration. In the ocean there are many heavy seas — the human heart has still heavier ones. There passed many thoughts, strong and weak mingled together, through Jörgen's head and heart, and he asked Elsé, —

"If Morten had a house as well as I, which of us two would you rather take?"

"But Morten has no house, and has no chance of getting one."

"But we think it is very likely he will have one."

"Oh! then I would take Morten, of course; but one can't live upon love."

And Jörgen reflected for the whole night over what

had passed. There was something in him he could not
himself account for; but he had one idea — it overpowered his love for Elsé, and it led him to Morten. What
he said and did there had been well considered by him
— he made his house over to Morten on the lowest possible terms, saying that he would himself prefer to go
into service. And Elsé kissed him in her gratitude
when she heard it, for she certainly loved Morten best.

At an early hour in the morning Jörgen was to take
his departure. The evening before, though it was already late, he fancied he would like to visit Morten once
more, so he went; and amongst the sand-hills he met
the old fisherman, who did not seem to think of his going
away, and who jested about all the girls being so much
in love with Morten. Jörgen cut him short, bade him
farewell, and proceeded to the house where Morten lived.
When he reached it he heard loud talking within : Morten was not alone. Jörgen was somewhat capricious.
Of all persons he would least wish to find Elsé there ;
and, on second thoughts, he would rather not give Morten
an opportunity of renewing his thanks, so he turned
back again.

Early next morning, before the dawn of day, he tied
up his bundle, took his provision box, and went down
from the sand-hills to the sea-beach. It was easier to
walk there than on the heavy sandy road; besides, it
was shorter, for he was first going to Fjaltring, near
Vosbjerg, where the eel-man lived, to whom he had
promised a visit.

The sea was smooth and beautifully blue, — shells of
different sorts lay around. These were the playthings of

his childhood — he now trod them under his feet. As he was walking along his nose began to bleed. That was only a trifle in itself, but it might have some meaning. A few large drops of blood fell .upon his arms; he washed them off, stopped the bleeding, and found that the loss of a little blood had actually made him feel lighter in his head and in his heart. A small quantity of sea-kale was growing in the sand; he broke a blade off of it, and stuck it in his hat. He tried to feel happy and confident now that he was going out into the wide world — "away from the door, a little way up the stream," as the eel's children had said; and the mother said, "Take care of bad men; they will catch you, skin you, cut you in pieces, and fry you." He repeated this to himself, and laughed at it. He would get through the world with a whole skin — no fear of that; for he had plenty of courage, and that was a good weapon of defence.

The sun was already high up, when, as he approached the small inlet between the German Ocean and Nissum-fiord, he happened to look back, and perceived at a considerable distance two people on horseback, and others following on foot: they were evidently making great haste, but it was nothing to him.

The ferry-boat lay on the other side of the narrow arm of the sea. Jörgen beckoned and called to the person who had charge of it. It came over and he entered it; but before he and the man who was rowing had got half way across, the men he had seen hurrying on reached the banks, and with threatening gestures shouted the name of the magistrate. Jörgen could not comprehend what they wanted, but considered it would be best to go back, and even took one of the oars to row the faster. The moment the boat neared the shore, people sprang into it, and before he had an idea of what they were going to do, they had thrown a rope round his hands, and made him their prisoner.

"Your evil deed will cost you your life," said they. "It is lucky we arrived in time to catch you."

It was neither more nor less than a murder he was accused of having committed. Morten had been found stabbed by a knife in his neck. One of the fishermen had, late the night before, met Jörgen going to the place where Morten lived. It was not the first time he had lifted a knife at him, they knew. He must be the murderer; therefore he must be taken into custody. Ringkjöbing was the most proper place to which to carry him, but it was a long way off. The wind was from the west. In less than half an hour they could cross the fiord at Skjærumaa, and from thence they had only a short way to go to Nörre-Vosborg, which was a strong place, with ramparts and moats. In the boat was a brother of the bailiff there, and he promised to obtain permission to put Jörgen for the present into the cell where Lange Margrethe had been confined before her execution.

Jörgen's defence of himself was not listened to; for a few drops of blood on his clothes spoke volumes against him. His innocence was clear to himself; and, if justice were not done him, he must give himself up to his fate.

They landed near the site of the old ramparts, where Sir Buggé's castle had stood, — there, where Jörgen, with his foster-father and mother, had passed on their way to

the funeral meeting, at which had been spent the four
brightest and pleasantest days of his childhood. He was
conveyed again the same way by the fields up to Nörre-
Vosborg, and yonder stood in full flower the elder tree,
and yonder the lindens shed their sweet perfume around ;
and he felt as if it had been only yesterday that he had
been there.

In the west wing of the castle is a subterranean passage under the high stairs; this leads to a low, vaulted
cell, in which Lange Margrethe had been imprisoned,
and whence she had been taken to the place of execution.
She had eaten the hearts of five children, and believed
that, could she have added two more to the number, she
would have been able to fly and to render herself·invis-
ible. In the wall there was a small, narrow air-hole.
No glass was in this rude window ; yet the sweetly-
scented linden tree on the outside could not send the
slightest portion of its refreshing perfume into that close,
mouldy dungeon. There was only a miserable pallet
there ; but a good conscience is a good pillow, therefore
Jörgen could sleep soundly.

The thick wooden door was locked, and it was further
secured by an iron bolt; but the nightmare of superstition can creep through a key-hole in the baronial castle
as in the fisherman's hut. It stole in where Jörgen was
sitting and thinking upon Lange Margrethe and her
misdeeds. Her last thoughts had filled that little room
the night before her execution; he remembered all the
magic that, in the olden times, was practised when the
lord of the manor, Svanwedel, lived there ; and it was
well known how, even now, the chained dog that stood on

the bridge was found every morning hung over the railing in his chain. All these tales recurred to Jörgen's mind, and made him shiver; and there was but one sun ray which shone upon him, and that was the recollection of the blooming elder and linden trees.

He would not be kept long here; he would be removed to Ringkjöbing, where the prison was equally strong.

These times were not like ours. It went hard with the poor then; for then it had not come to pass that peasants found their way up to lordly mansions, and that from these regiments coachmen and other servants became judges in the petty courts, which were invested with the power to condemn, for perhaps a trifling fault, the poor man to be deprived of all his goods and chattels, or to be flogged at the whipping-post. A few of these courts still remain; and in Jutland, far from "the King's Copenhagen," and the enlightened and liberal government, even now the law is not always very wisely administered: it certainly was not so in the case of poor Jörgen.

It was bitterly cold in the place where he was confined. When was this imprisonment to be at an end? Though innocent, he had been cast into wretchedness and solitude — that was his fate. How things had been ordained for him in this world, he had now time to think over. Why had he been thus treated, — his portion made so hard to bear? Well, this would be revealed "in that other life" which assuredly awaits all. In the humble cottage that belief had been engrafted into him, which, amidst the grandeur and brightness of his Spanish home, had never shone upon his father's heart: *that* now, in the midst of cold and darkness, became his consolation, God's gift of grace, which never can deceive.

The storms of spring were now raging; the roaring of the German Ocean was heard far inland; but just when the tempest had lulled, it sounded as if hundreds of heavy wagons were driving over a hard tunnelled road. Jörgen heard it even in his dungeon, and it was a change in the monotony of his existence. No old melody could have gone more deeply to his heart than these sounds — the rolling ocean — the free ocean — on which one can be borne throughout the world, fly with the wind, and wherever one went have one's own house with one, as the snail has his — to stand always upon home's ground, even in a foreign land.

How eagerly he listened to the deep rolling! How remembrances hurried through his mind! "Free — free — how delightful to be free, even without soles to one's shoes, and in a coarse patched garment!" The very idea brought the warm blood rushing into his cheeks, and he struck the wall with his fist in his vain impatience. Weeks, months, a whole year had elapsed, when a gipsy named Niels Tyv — "the horse-dealer," as he was also called — was arrested, and then came better times: it was ascertained what injustice had been done to Jörgen.

To the north of Ringkjöbing Fiord, at a small country inn, on the evening of the day previous to Jörgen's leaving home, and the committal of the murder, Niels Tyv and Morten had met each other. They drank a little together, not enough certainly to get into any man's head, but enough to set Morten talking too freely. He went on chattering, as he was fond of doing, and he mentioned

2 *   C

that he had bought a house and some ground, and was going to be married. Niels thereupon asked him where was the money which was to pay it, and Morten struck his pocket pompously, exclaiming in a vaunting manner, —

" Here, where it should be ! "

That foolish bragging answer cost him his life; for when he left the little inn Niels followed him, and stabbed him in the neck with his knife, in order to rob him of the money, which, after all, was not to be found.

There was a long trial and much deliberation : it is enough for us to know that Jörgen was set free at last. But what compensation was made to him for all he had suffered that long weary year in a cold, gloomy prison, secluded from all mankind? Why, he was assured that it was fortunate he was innocent, and he might now go about his business! The burgomaster gave him ten marks for his travelling expenses, and several of the townspeople gave him ale and food. They were very good people. Not all, then, would " skin you, and lay you on the frying-pan ! " But the best of all was that the trader Brönne from Skagen, he to whom, a year before, Jörgen intended to have hired himself, was just at the time of his liberation on business at Ringkjöbing. He heard the whole story ; he had a heart and understanding; and, knowing what Jörgen must have suffered and felt, he was determined to do what he could to improve his situation, and let him see that there were some kind-hearted people in the world.

From a jail to freedom — from solitude and misery to a home which, by comparison, might be called a heaven —

to kindness and love, he now passed. This also was to be a trial of his character. No chalice of life is altogether wormwood. A good person would not fill such for a child: would, then, the Almighty Father, who is all love, do so?

"Let all that has taken place be now buried and forgotten," said the worthy Mr. Brönne. "We shall draw a thick line over last year. We shall burn the almanac. In two days we shall start for that blessed, peaceful, pleasant Skagen. It is said to be only a little insignificant nook in the country; but a nice warm nook it is, with windows open to the wide world."

That *was* a journey — that *was* to breathe the fresh air again — to come from the cold, damp prison-cell out into the warm sunshine!

The heather was blooming on the moorlands; the shepherd boys sat on the tumuli and played their flutes, which were manufactured out of the bones of sheep; the FATA MORGANA, the beautiful mirage of the desert, with its hanging seas and undulating woods, showed itself; and that bright, wonderful phenomenon in the air, which is called the "Lokéman driving his sheep."

Towards Limfiorden they passed over the Vandal's land; and towards Skagen they journeyed where the men with the long beards, *Langbarderne,** came from.

---

* Langobarder, a northern tribe, which, in very ancient times, dwelt in the north of Jutland. From thence they migrated to the north of Germany, where, according to Tacitus, they lived about the period of the birth of Christ, and were a poor but brave people. Their original name was Vinuler, or Viniler. "When these Viniler," say the traditions, or rather fables of Scandinavia, "were at war with the

In that locality it was that, during the famine under King
Snio, all old people and young children were ordered to
be put to death; but the noble lady, Gambaruk, who
was the heiress of that part of the country, insisted that
the children should rather be sent out of the country.
Jörgen was learned enough to know all about this; and,
though he was not acquainted with the Langobarders'
country beyond the lofty Alps, he had a good idea what
it must be, as he had himself, when a boy, been in the
south of Europe, in Spain.   Well did he remember the
heaped up piles of fruit, the red pomegranate flowers, the
din, the clamour, the tolling of bells in the Spanish city's
great hive; but all was more charming at home, and
Denmark was Jörgen's home.

At length they reached Vendilskaga, as Skagen is
called in the old Norse and Icelandic writings.   For
miles and miles, interspersed with sand-hills and cultivated land, houses, farms, and drifting sand-banks,
stretched, and stretch still, towards Gammel-Skagen,

Vandals, and the latter went to Odin to beseech him to grant them
the victory, and received for answer that Odin would award the victory to those whom he beheld first at sunrise, the warlike female,
Gambaruk, or Gunborg, who was mother to the leaders of the Viniler
— Ebbe and Aage — applied to Frigg, Odin's wife, to entreat victory
for her people.   The goddess advised that the females of the tribe
should let down their long hair so as to imitate beards, and, early in
the morning, should stand with their husbands in the east, where Odin
would look out.   When, at sunrise, Odin saw them, he exclaimed,
' Who are these long-bearded people?' whereupon Frigga replied, that,
since he had bestowed a name upon them, he must also give them the
victory.   This was the origin of the *Longobardi*, who, after many wan-
derings, found their way into Italy, and, under ALBOIN, founded the
kingdom of Lombardy." — *Trans.*

Wester and Osterby, out to the lighthouse near Grenen, a waste, a desert, where the wind drives before it the loose sand, and where sea-gulls and wild swans send forth their discordant cries in concert. To the south-west, a few miles from Grenen, lies High, or Old Skagen, where the worthy Brönne lived, and where Jörgen was also to reside. The house was tarred, the small out-houses had each an inverted boat for a roof. Pieces of wrecks were knocked up together to form pigsties. Fences there were none, for there was nothing to in-close; but upon cords, stretched in long rows one over the other, hung fish cut open, and drying in the wind. The whole beach was covered with heaps of putrefying herrings; nets were scarcely ever thrown into the water, for the herrings were taken in loads on the land. There was so vast a supply of this sort of fish that people either threw them back into the sea, or left them to rot on the sands.

The trader's wife and daughter — indeed, the whole household — came out rejoicing to meet the father of the family when he returned home. There was such a shak-ing of hands — such exclamations and questions! And what a charming countenance and beautiful eyes the daughter had!

The interior of the house was large and extremely comfortable. Various dishes of fish were placed upon the table; among others some delicious plaice, which might have been a treat for a king; wine from Skagen's vineyard — the vast ocean — from which the juice of the grape was brought on shore both in casks and bottles.

When the mother and daughter afterwards heard who
Jörgen was, and how harshly he had been treated, though
innocent of all crime, they looked very. kindly at him ;
and most sympathising was the expression of the daughter's eyes, the lovely Miss Clara.   Jörgen found a happy
home at Gammel-Skagen.   It did his heart good, and
the poor young man had suffered much, even the bitter-
ness of unrequited love, which either hardens or softens
the heart. Jörgen's was soft enough now; there was a
vacant place within it, and he was still so young.

It was, perhaps, fortunate that in about three weeks
Miss Clara was going in one of her father's ships up to
Christiansand, in Norway, to visit an aunt, and remain
there the whole winter.   The Sunday before her departure they all went to church together, intending to partake
of the sacrament.   It was a large, handsome church, and
had several hundred years before been built by the
Scotch and Dutch a little way from where the town was
now situated.   It had become somewhat dilapidated, was
difficult of access, the way to it being through deep,
heavy sand; but the disagreeables of the road were
willingly encountered in order to enter the house of God
— to pray, sing psalms, and hear a sermon there.   The
sand was, as it were, banked up against, and even higher
than, the circular wall of the churchyard ; but the graves
therein were kept carefully free of the drifting sand.

This was the largest church to the north of Limfiorden.   The Virgin Mary, with a crown of gold on her
head, and the infant Jesus in her arms, stood as if in life
in the altar-piece ; the holy apostles were carved on the
chancel; and on the walls above were to be seen the

portraits of the old burgomasters and magistrates of Skagen, with their insignia of office: the pulpit was richly carved. The sun was shining brightly into the church, and glancing on the crown of brass and the little ship that hung from the roof.

Jörgen felt overcome by a kind of childish feeling of awe, mingled with reverence, such as he had experienced when as a boy he had stood within the magnificent Spanish cathedral; but he knew that here his feelings were shared by many. After the sermon the sacrament was administered. Like the others, he tasted the consecrated bread and wine, and he found that he was kneeling by the side of Miss Clara; but he was so much absorbed in his devotions, and in the sacred rite, that it was only when about to rise that he observed who was his immediate neighbour, and perceived that tears were streaming down her cheeks.

Two days after this she sailed for Norway, and Jörgen made himself useful on the farm, and at the fishery, in which there was much more done then than is now-a-days. The shoals of mackerel glittered in the dark nights, and showed the course they were taking; the crabs gave piteous cries when pursued, for fishes are not so mute as they are said to be. Every Sunday when he went to church, and gazed on the picture of the Virgin in the altar-piece, Jörgen's eyes always wandered to the spot where Clara had knelt by his side; and he thought of her, and how kind she had been to him.

Autumn came, with its hail and sleet; the water washed up to the very town of Skagen; the sand could not absorb all the water, so that people had to wade through

it. The tempests drove vessel after vessel on the fatal
reefs; there were snow storms and sand storms; the
sand drifted against the houses, and closed up the en-
trances in some places, so that people had to creep out
by the chimneys; but that was nothing remarkable up
there. While all was thus bleak and wretched without,
within there were warmth and comfort. The mingled
peat and wood fires — the wood obtained from wrecked
ships — crackled and blazed cheerfully, and Mr. Brönne
read aloud old chronicles and legends; among others, the
story of Prince Hamlet of Denmark, who, coming from
England, landed near Bovbjerg, and fought a battle there.
His grave was at Ramme, only a few miles from the
place where the eel-man lived. Hundreds of tumuli, the
graves of the giants and heroes of old, were still visible
all over the wide heath — a great churchyard. Mr.
Brönne had himself been there, and had seen Hamlet's
grave. They talked of the olden times — of their neighbours, the English and Scotch; and Jörgen sang the ballad about "The King of England's Son" — about the
splendid ship — how it was fitted up:—

> " How on the gilded panels stood
> Engraved our Lord's commandments good;
>   *   *   *   *   *
> And, clasping a sweet maiden, how
> The prince stood sculptured on the prow! "

Jörgen sang these lines in particular with much empha-
sis, whilst his dark eyes sparkled; but his eyes had
always been bright from his earliest infancy.

There were songs, and readings, and conversation, and
everything to make the winter season pass as pleasantly

as possible; there was prosperity in the house, plenty of comfort for the family, and plenty even for the lowest animals on the property; the shelves shone with rows of bright, well-scoured pewter plates and dishes; and from the roof hung sausages and hams, and other winter stores in abundance. Such may be seen even now in the many rich farm-houses on the west coast — the same evidences of plenty, the same comfortable rooms, the same good-humour, the same, and perhaps a little more, information. Hospitality reigns there as in an Arab's tent.

Jörgen had never before spent his time so happily since the pleasant days of his childhood at the funeral feast; and yet Miss Clara was absent — present only in thought and conversation.

In April a vessel was going up to Norway, and Jörgen was to go in it. He was in high spirits, and, according to Mrs. Brönne, he was so lively and good-humoured, it was quite a pleasure to see him.

"And it is quite a pleasure to see you also," said her husband. "Jörgen has enlivened all our winter evenings, and you with them; you have become young again, and really look quite handsome. You were formerly the prettiest girl in Viborg, and that is saying a great deal, for I have always thought the girls prettier there than anywhere else."

Jörgen said nothing to this. Perhaps he did not believe that the Viborg girls were prettier than any others; at any rate, he was thinking of one from Skagen, and he was now about to join her. The vessel had a fair, fresh breeze; therefore he arrived at Christiansand in half a day.

Early one morning the trader, Mr. Brönne, went out
to the lighthouse that is situated at some distance from
Gammel-Skagen, and near Grenen. The signal-lights
had been extinguished for some time, for the sun had
risen tolerably high before he reached the tower. Away,
to some distance beyond the most remote point of land,
stretched the sand-banks under the water. Beyond
these, again, he perceived many ships, and among them
he thought he recognised, by aid of the spy-glass, the
" Karen Brönne," as his own vessel was called ; and he
was right. It was approaching the coast, and Clara and
Jörgen were on board. The Skagen lighthouse and the
spire of its church looked to them like a heron and a
swan upon the blue water. Clara sat by the gunwale,
and saw the sand-hills becoming little by little more and
more apparent. If the wind only held fair, in less than
an hour they would reach home ; so near were they to
happiness, and yet, alas ! how near to death !

A plank sprung in the ship. The water rushed in.
They stopped it as well as they could, and used the
pumps vigorously. All sail was set, and the flag of distress was hoisted. They were about a Danish mile off.
Fishing-boats were to be seen, but were far away. The
wind was fair for them. The current was also in their
favour, but not strong enough. The vessel sank. Jörgen
threw his right arm around Clara.

With what a speaking look did she not gaze into his
eyes when, imploring Our Lord for help, he threw himself with her into the sea ! She uttered one shriek, but
she was safe. He would not let her slip from his grasp.
The words of the old ballad, —

" And, clasping a sweet maiden, how
The prince stood sculptured on the prow," —

were now carried into effect by Jörgen in that agonising
hour of danger and deep anxiety. He felt the advantage of being a good swimmer, and exerted himself to
the utmost with his feet and one hand; the other was
holding fast the young girl. Every possible effort he
made to keep up his strength in order to reach the land.
He heard Clara sigh, and perceived that a kind of con-
vulsive shuddering had seized her; and he held her the
tighter. A single heavy wave broke over them — the
current lifted them. The water was so clear, though
deep, that Jörgen thought for a moment he could see the
shoals of mackerel beneath; or was it Leviathan himself
who was waiting to swallow them? The clouds cast a
shadow over the water, then again came the dancing
sunbeams; harshly-screaming birds, in flocks, wheeled
over him; and the wild ducks that, heavy and sleepy,
allow themselves to drive on with the waves, flew up in
alarm from before the swimmer. He felt that his strength
was failing; but the shore was close at hand, and help
was coming, for a boat was near. Just then he saw distinctly under the water a white, staring figure; a wave
lifted him, the figure came nearer, he felt a violent blow,
it became night before his eyes — all had disappeared
for him.

There lay, partially imbedded in the sand-bank, the
wreck of a ship; the sea rolled over it, but the white
figure-head was supported by an anchor, the sharp iron
of which stuck up almost to the surface of the water. It
was against this that Jörgen had struck himself when the

current had driven him forward with sudden force. Stunned and fainting, he sank with his burden, but the succeeding wave threw him and the young girl up again.

The fishermen had now reached them, and they were taken into the boat. Blood was streaming over Jörgen's face; he looked as if he were dead, but he still held the girl in so tight a grasp that it was with the utmost difficulty she could be wrenched from his encircling arm. As pale as death, and quite insensible, she lay at full length at the bottom of the boat, which steered towards Skagen.

All possible means were tried to restore Clara to animation, but in vain — the poor young woman was dead. Long had Jörgen been buffeting the waves with a corpse — exerting his utmost strength and straining every nerve for a dead body.

Jörgen still breathed; he was carried to the nearest house on the inner side of the sand-hills. A sort of army surgeon who happened to be at the place, who also acted in the capacities of smith and huckster, attended him until the next day, when a physician from Hjörring, who had been sent for, arrived.

The patient was severely wounded in the head, and suffering from a brain fever. For a time he uttered fearful shrieks, but on the third day he sank into a state of drowsiness, and his life seemed to hang upon a thread: that *it* might snap, the physician said, was the best that could be wished for Jörgen.

" Let us pray our Lord that he may be taken; he will never more be a rational man."

But he was not taken; the thread of life would not

break, though memory was swept away, and all the powers and faculties of his mind were gone. It was a frightful change. A living body was left — a body that was
to regain health and go about again.

Jörgen remained in the trader Brönne's house.

"He was brought into this lamentable condition by his
efforts to save our child," said the old man; "he is now
our son."

Jörgen was called "an idiot;" but that was a term not
exactly applicable to him. He was like a musical instrument, the strings of which are loose, and can no longer,
therefore, be made to sound. Only once, for a few minutes, they seemed to resume their elasticity, and they
vibrated again. Old melodies were played, and played
in time. Old images seemed to start up before him.
They vanished — all glimmering of reason vanished, and
he sat again staring vacantly around, without thought,
without mind. It was to be hoped that he did not suffer
anything. His dark eyes had lost their intelligence;
they looked only like black glass that could move
about.

Everybody was sorry for the poor idiot Jörgen.

It was he who, before he saw the light of day, was
destined to a career of earthly prosperity, of wealth and
happiness, so great that it was "*frightful pride, overween-
ing arrogance,*" to wish for, or to believe in, a future life!
All the high powers of his soul were wasted. Nothing
but hardships, sufferings, and disappointments had been
dealt out to him. A valuable bulb he was, torn up from
his rich native soil, and cast upon distant sands to rot and
perish. Was that being, made in the image of God,

worth nothing more? Was he but the sport of accidents
or of chance? No! The God of infinite love would
give him a portion in another life for what he had suffered and been deprived of here.

"The Lord is good to all: and His tender mercies are
over all His works."

These consolatory words, from one of the Psalms of
David, were repeated in devout faith by the pious old
wife of the trader Brönne; and her heartfelt prayer was,
that our Lord would soon release the poor benighted being, and receive him into God's gift of grace — everlasting life.

In the churchyard, where the sand had drifted into
piles against the walls, was Clara buried. It appeared
as if Jörgen had never thought about her grave; it did
not enter into the narrow circle of his ideas, which now
only dwelt among wrecks of the past. Every Sunday
he accompanied the family to church, and he generally
sat quiet with a totally vacant look; but one day, while
a psalm was being sung, he breathed a sigh, his eyes
lightened up, he turned them towards the altar — towards
that spot where, more than a year before, he had knelt,
with his dead friend at his side. He uttered her name,
became as white as a sheet, and tears rolled down his
cheeks.

He was helped out of church, and then he said that he
felt quite well, and did not think anything had been the
matter with him; the short flash of memory had already
faded away from him — the much-tried, the sorely-smit-
ten of God. Yet that God, our Creator, is all wisdom

and all love, who can doubt? Our hearts and our reason acknowledge it, and the Bible proclaims it. "His tender mercies are over all His works."

In Spain, where, amidst laurels and orange trees, the Moorish golden cupolas glitter in the warm air, where songs and castanets are heard, sat, in a splendid mansion, a childless old man. Children were passing through the streets in a procession, with lights and waving banners. How much of his enormous wealth would he not have given to possess one child — to have had spared to him his daughter and her little one, who perhaps never beheld the light of day in this world. If so, how would it behold the light of eternity — of paradise? "Poor, poor child!"

Yes; poor child — nothing but a child — and yet in his thirtieth year! for to such an age had Jörgen attained there in Gammel-Skagen.

The sand-drifts had found their way even over the graves in the churchyard, and up to the very walls of the church itself; yet here, amidst those who had gone before them — amidst relatives and friends — the dead were still buried. The good old Brönne and his wife reposed there, near their daughter, under the white sand.

It was late in the year — the time of storms; the sand-hills smoked, the waves rolled mountains high on the raging sea; the birds in hosts, like dark tempestuous clouds, passed screeching over the sand-hills; ship after ship went ashore on the terrible reefs between Skagen's Green and Huusby-Klitter.

One afternoon Jörgen was sitting alone in the parlour, and suddenly there rushed upon his shattered mind a feeling akin to the restlessness which so often, in his younger years, had driven him out among the sand-hills, or upon the heath.

"Home! home!" he exclaimed. No one heard him. He left the house, and took his way to the sand-hills. The sand and the small stones dashed against his face, and whirled around him.   He went towards the church; the sand was lying banked up against the walls, and half way up the windows; but the walk up to the church was freer of it.   The church door was not locked, it opened easily, and Jörgen entered the sacred edifice.

The wind went howling over the town of Skagen; it was blowing a perfect hurricane, such as had not been known in the memory of the oldest man living — it was most fearful weather.   But Jörgen was in God's house, and while dark night came on around him, all seemed light within; it was the light of the immortal soul which is never to be extinguished. He felt as if a heavy stone had fallen from his head; he fancied that he heard the organ playing, but the sounds were those of the storm and the roaring sea.   He placed himself in one of the pews, and he fancied that the candles were lighted one after the other, until there was a blaze of brilliancy such as he had beheld in the cathedral in Spain; and all the portraits of the old magistrates and burgomasters became imbued with life, descended from the frames in which they had stood for years, and placed themselves in the choir.   The gates and side doors of the church opened, he thought, and in walked all the dead, clothed in the grandest costumes of their times, whilst music floated in the air; and when they had seated themselves in the different pews, a solemn hymn arose, and swelled like the rolling of the sea.

Among those who had joined the spirit throng were his old foster-father and mother from Huusby-Klitter, and his kind friend Brönne and his wife; and at their side, but close to himself, sat their mild, lovely daughter. She held out her hand to him, Jörgen thought, and they went up to the altar where once they had knelt together; the priest joined their hands, and pronounced those words and that blessing which were to hallow for them life and love. Then music's tones pealed around — the organ, wind instruments, and voices combined — until there arose a volume of sound sufficient to shake the very tombstones over the graves.

Presently the little ship that hung under the roof moved towards him and Clara. It became large and magnificent, with silken sails and gilded masts; the anchor was of the brightest gold, and every rope was of silk cord, as described in the old song. He and his bride stepped on board, then the whole multitude in the church followed them, and there was room for all. He fancied that the walls and vaulted roof of the church turned into blooming elder and linden trees, which diffused a sweet perfume around. It was all one mass of verdure. The trees bowed themselves, and left an open space; then the ship ascended gently, and sailed out through the air above the sea. Every light in the church looked like a star. The wind commenced a hymn, and all sang with it: "In love to glory!" "No life shall be lost!" "Away to supreme happiness!" " Hallelujah!"

These words were his last in this world. The cord

3   D

had burst which held the dying soul. There lay but a
cold corpse in the dark church, around which the storm
was howling, and which it was overwhelming with the
drifting sand.

The next morning was a Sunday; the congregation
and their pastor came at the hour of church service. The
approach to the church had been almost impassable on
account of the depth of the sand, and when at length
they reached it, they found an immense sand-heap piled
up before the door of the church — the drifting sand had
closed up all entrance to its interior. The clergyman
read a prayer, and then said that, as God had locked the
doors of that holy house, they must go elsewhere and
erect another for His service.

They sang a psalm, and retired to their homes.

Jörgen could not be found either at Skagen or amidst
the sand-hills, where every search was made for him. It
was supposed that the wild waves, which had rolled so
far up on the sands, had swept him off.

But his body lay entombed in a large sarcophagus —
in the church itself. During the storm God had cast
earth upon his coffin — heavy piles of quicksand had
accumulated there, and lie there even now.

The sand had covered the lofty arches, sand-thorns
and wild roses grow over the church, where the wayfarer
now struggles on towards its spire, which towers above
the sand, an imposing tombstone over the grave, seen
from miles around — no king had ever a grander one!
None disturb the repose of the dead — none knew where
Jörgen lay, until now — the storm sang the secret for me
among the sand-hills!

The Mud-king's Daughter.

THE storks are in the habit of relating to their little ones many tales, all from the swamps and the bogs. They are, in general, suitable to the ages and comprehensions of the hearers. The smallest youngsters are contented with mere sound, such as "krible, krable, plurre-murre." They think that wonderful; but the more advanced require something rational, or at least something about their family. Of the two most ancient and longest traditions that have been handed down among the storks, we are all acquainted with one — that about Moses, who was placed by his mother on the banks of the Nile, was found there by the king's daughter, was well brought up, and became a great man, such as has never been heard of since in the place where he was buried.

The other story is not well known, probably because it is a tale of home; yet it has passed down from one stork grandam to another for a thousand years, and each succeeding narrator has told it better and better, and now we shall tell it best of all.

The first pair of storks who related this tale had them-

selves something to do with its events. The place of their
summer sojourn was at the Viking's loghouse, up by
the wild morass, at Vendsyssel.   It is in Hjöring district,
away near Skagen, in the north of Jutland, speaking with
geographical precision.   It is now an enormous bog, and
an account of it can be read in descriptions of the country.   This place was once the bottom of the sea; but the
waters have receded, and the ground has risen.   It
stretches itself for miles on all sides, surrounded by wet
meadows and pools of water, by peat-bogs, cloudberries,
and miserable stunted trees.   A heavy mist almost
always hangs over this place, and about seventy years
ago wolves were found there.   It is rightly called, " the
wild morass; " and one may imagine how savage it must
have been, and how much swamp and sea must have
existed there a thousand years ago.   Yes, in these
respects the same was to be seen there as is to be seen
now.   The rushes had the same height, the same sort of
long leaves, and blue-brown, feather-like flowers, that
they bear now; the birch tree stood with its white bark,
and delicate drooping leaves, as now; and, in regard to
the living creatures, the flies had the same sort of crape
clothing as they wear now; and the storks' bodies were
white, with black and red stockings.   Mankind, on the
contrary, at that time wore coats cut in another fashion
from what they do in our days; but every one of them,
serf or huntsman, whosoever he might be who trod upon
the quagmire, fared a thousand years ago as they fare
now: one step forward — they fell in, and sank down to
the MUD-KING, as *he* was called who reigned below in
the great morass kingdom.   Very little is known about
his government; but that is, perhaps, a good thing.

Near the bog, close by Liimfjorden, lay the Viking's loghouse of three stories high, and with a tower and stone cellars. The storks had built their nest upon the roof of this dwelling. The female stork sat upon her eggs, and felt certain they would be all hatched.

One evening the male stork remained out very long, and when he came home he looked rumpled and flurried.

"I have something very terrible to tell thee," he said to the female stork.

"Thou hadst better keep it to thyself," said she. "Re-member I am sitting upon the eggs: a fright might do me harm, and the eggs might be injured."

"But it *must* be told thee," he replied. "She has come here — the daughter of our host in Egypt. She has ventured the long journey up hither, and she is lost."

"She who is of the fairies' race? Speak, then! Thou knowest that I cannot bear suspense while I am sitting."

"Know, then, that she believed what the doctors said, which thou didst relate to me. She believed that the bog-plants up here could cure her invalid father; and she has flown hither, in the magic disguise of a swan, with the two other swan princesses, who every year come hither to the north to bathe and renew their youth. She has come, and she is lost."

"Thou dost spin the matter out so long," muttered the female stork, "the eggs will be quite cooled. I cannot bear suspense just now."

"I will come to the point," replied the male. "This evening I went to the rushes where the quagmire could bear me. Then came three swans. There was some-thing in their motions which said to me, 'Take care;

they are not real swans; they are only the appearance of swans, created by magic.' Thou wouldst have known as well as I that they were not of the right sort."

"Yes, surely," she said; "but tell me about the princess. I am tired of hearing about the swans.

"In the midst of the morass — here, I must tell thee, it is like a lake," said the · male stork — "thou canst see a portion of it if thou wilt raise thyself up a moment — yonder, by the rushes and the green morass, lay a large stump of an alder tree. The three swans alighted upon it, flapped their wings, and looked about them. One of them cast off her swan disguise, and I recognised in her our royal princess from Egypt. She sat now with no other mantle around her than her long dark hair. I heard her desire the other two to take good care of her magic swan garb, while she ducked down under the water to pluck the flower which she thought she saw. They nodded, and raised the empty feather dress between them. 'What are they going to do with it?' said I to myself; and she probably asked herself the same question. The answer came too soon, for I saw them take flight up into the air with her charmed feather dress. 'Dive thou there!' they cried. 'Never more shalt thou fly in the form of a magic swan — never more shalt thou behold the land of Egypt. Dwell thou *in the wild morass!*' And they tore her magic disguise into a hundred pieces, so that the feathers whirled around about as if there were a fall of snow; and away flew the two worthless princesses."

"It is shocking!" said the lady stork; "I can't bear to hear it. Tell me what more happened."

"The princess sobbed and wept. Her tears trickled down upon the trunk of the alder tree, and then it moved; for it was the mud-king himself—he who dwells in the morass. I saw the trunk turn itself, and then there was no more trunk—it struck up two long miry branches like arms; then the poor child became dreadfully alarmed, and she sprang aside upon the green slimy coating of the marsh; but it could not bear me, much less her, and she sank immediately in. The trunk of the alder tree went down with her— it was that which had dragged her down: then arose to the surface large black bubbles, and all further traces of her disappeared. She is now buried in 'the wild morass;' and never, never shall she return to Egypt with the flower she sought. Thou couldst not have borne to have seen all this, mother."

"Thou hadst no business to tell me such a startling tale at a time like this. The eggs may suffer. The princess can take care of herself: she will no doubt be rescued. If it had been me or thee, or any of our family, it would have been all over with us."

"I will look after her every day, however," said the male stork; and so he did.

A long time had elapsed, when one day he saw that far down from the bottom was shooting up a green stem, and when it reached the surface a leaf grew on it. The leaf became broader and broader; close by it came a bud; and one morning, when the stork flew over it, the bud opened in the warm sunshine, and in the centre of it lay a beautiful infant, a little girl, just as if she had been taken out of a bath. She so strongly resembled the princess from Egypt, that the stork at first thought it was herself who had become an infant again; but when he considered the matter he came to the conclusion that she was the daughter of the princess and the mud-king, therefore she lay in the calyx of a water-lily.

"She cannot be left lying there," said the stork to himself; "yet in my nest we are already too over-crowded. But a thought strikes me. The Viking's wife has no children; she has much wished to have a pet. I am often blamed for bringing little ones. I shall now, for once, do so in reality. I shall fly with this infant to the Viking's wife: it will be a great pleasure to her."

And the stork took the little girl, flew to the loghouse, knocked with his beak a hole in the window-pane of stretched bladder, laid the infant in the arms of the Viking's wife, then flew to his mate, and unburdened his mind to her; while the little ones listened attentively, for they were old enough now to do that.

"Only think, the princess is not dead. She has sent her little one up here, and now it is well provided for."

"I told thee from the beginning it would be all well," said the mother stork. "Turn thy thoughts now to thine own family. It is almost time for our long journey; I begin now to tingle under the wings. The cuckoo and the nightingale are already gone, and I hear the quails saying that we shall soon have a fair wind. Our young ones are quite able to go, I know that."

How happy the Viking's wife was when, in the morning, she awoke and found the lovely little child lying on her breast! She kissed it and caressed it, but it screeched frightfully, and floundered about with its little arms and legs: IT evidently seemed little pleased. At last it cried itself to sleep, and as it lay there it was one of the most beautiful little creatures that could be seen. The Viking's wife was so pleased and happy, she took it into her head that her husband, with all his retainers, would come as unexpectedly as the little one had done; and she set herself and the whole household to work, in order that everything might be ready for their reception. The coloured tapestry which she and her women had embroidered with representations of their gods — ODIN, THOR, and FREIA, as they were called — were hung up; the serfs were ordered to clean and polish the old shields with which the walls were to be decorated; cushions were laid on the benches; and dry logs of wood were heaped on the fireplace in the centre of the hall, so that the pile might be easily lighted. The Viking's wife laboured so hard herself that she was quite tired by the evening, and slept soundly.

When she awoke towards morning she became much alarmed, for the little child was gone. She sprang up, lighted a twig of the pine tree, and looked about; and, to her amazement, she saw, in the part of the bed to which she stretched her feet, not the beautiful infant, but a great ugly frog. She was so much disgusted with it that she took up a heavy stick, and was going to kill the nasty creature; but it looked at her with such wonderfully sad and speaking eyes that she could not strike it. Again she searched about. The frog gave a faint, pitiable cry. She started up, and sprang from the bed to the

3*

window; she opened the shutters, and at the same moment the sun streamed in, and cast its bright beams upon
the bed and upon the large frog; and all at once it seemed
as if the broad mouth of the noxious animal drew itself
in, and became small and red — the limbs stretched themselves into the most beautiful form — it was her own little lovely child that lay there, and no ugly frog.

"What is all this?" she exclaimed. "Have I dreamed
a bad dream? That certainly is my pretty little elfin
child lying yonder." And she kissed it and strained it
affectionately to her heart; but it struggled, and tried to
bite like the kitten of a wild cat.

Neither the next day nor the day after came the Viking, though he was on the way, but the wind was against
him; it was for the storks. A fair wind for one is a contrary wind for another.

In the course of a few days and nights it became evident to the Viking's wife how things stood with the little
child — that it was under the influence of some terrible
witchcraft. By day it was as beautiful as an angel, but
it had a wild, evil disposition; by night, on the contrary,
it was an ugly frog, quiet, except for its croaking, and
with melancholy eyes. It had two natures, that changed
about, both without and within. This arose from the little girl whom the stork had brought possessing by day
her own mother's external appearance, and at the same
time her father's temper; while by night, on the contrary,
she showed her connection with him outwardly in her
form, whilst her mother's mind and heart inwardly became hers. What art could release her from the power
which exercised such sorcery over her? The Viking's

wife felt much anxiety and distress about it, and yet her heart hung on the poor little being, of whose strange state she thought she should not dare to inform her husband when he came home; for he assuredly, as was the custom, would put the poor child out on the high road, and let any one take it who would. The Viking's good-natured wife had not the heart to allow this; therefore she resolved that he should never see the child but by day.

At dawn of day the wings of the storks were heard fluttering over the roof. During the night more than a hundred pairs of storks had been making their preparations, and now they flew up to wend their way to the south.

"Let all the males be ready," was the cry. "Let their mates and little ones join them."

"How light we feel!" said the young storks, who were all impatience to be off. "How charming to be able to travel to other lands!"

"Keep ye all together in one flock," cried the father and mother, "and don't chatter so much — it will take away your breath."

So they all flew away.

About the same time the blast of a horn sounding over the heath gave notice that the Viking had landed with all his men; they were returning home with rich booty from the Gallic coast, where the people, as in Britain, sang in their terror, —

" Save us from the savage Normands! "

What life and bustle were now apparent in the Viking's castle near "the wild morass"! Casks of mead were

brought into the hall, the pile of wood was lighted, and
horses were slaughtered for the grand feast which was to
be prepared.   The sacrificial priests sprinkled with the
horses' warm blood the slaves who were to assist in the
offering.   The fires crackled, the smoke rolled up under
the roof, the soot dropped from the beams; but people
were accustomed to that.   Guests were invited, and they
brought handsome gifts; rancour and falseness were forgotten — they all became drunk together, and they thrust
their doubled fists into each other's faces — which was a
sign of good-humour.   The skald — he was a sort of
poet and musician, but at the same time a warrior — who
had been with them, and had witnessed what he sang
about, gave them a song, wherein they heard recounted
all their achievements in battle, and wonderful adventures.   At the end of every verse came the same refrain, —

" Fortune dies, friends die, one dies one's self; but a glorious name
never dies."

And then they all struck on their shields, and thundered
with their knives or their knuckle-bones on the table, so
that they made a tremendous noise.

The Viking's wife sat on the cross bench in the open
banquet hall.   She wore a silk dress, gold bracelets, and
large amber beads.   She was in her grandest attire, and
the skald named her also in his song, and spoke of the
golden treasure she had brought her husband; and HE
rejoiced in the lovely child he had only seen by daylight,
in all its wondrous beauty.   The fierce temper which
accompanied her exterior charms pleased him.   " She
might become," he said, " a stalwart female warrior, and

able to kill a giant adversary." She never even blinked her eyes when a practised hand, in sport, cut off her eyebrows with a sharp sword.

The mead casks were emptied, others were brought up, and these, too, were drained; for there were folks present who could stand a good deal. To them might have been applied the old proverb, "The cattle know when to leave the pasture; but an unwise man never knows the depth of his stomach."

Yes, they all knew it; but people often know the right thing, and do the wrong. They knew also that "one wears out one's welcome when one stays too long in another man's house;" but they remained there for all that. Meat and mead are good things. All went on merrily, and towards night the slaves slept amidst the warm ashes, and dipped their fingers into the fat skimmings of the soup, and licked them. It was a rare time!

And again the Viking went forth on an expedition, notwithstanding the stormy weather. He went after the crops were gathered in. He went with his men to the coast of Britain — "it was only across the water," he said — and his wife remained at home with her little girl; and it was soon to be seen that the foster-mother cared almost more for the poor frog, with the honest eyes and plaintive croaking, than for the beauty who scratched and bit everybody around.

The raw, damp, autumn mist, that loosens the leaves from the trees, lay over wood and hedge; "Birdfeatherless," as the snow is called, was falling thickly; winter was close at hand. The sparrows seized upon the storks' nest, and talked over, in their fashion, the absent owners. They themselves, the stork pair, with all their young ones, where were they now?

The storks were now in the land of Egypt, where the sun was shining warmly as with us on a lovely summer day. The tamarind and the acacia grew there; the moonbeams streamed over the temples of Mahomet. On the slender minarets sat many a pair of storks, reposing after their long journey; the whole immense flock had fixed themselves, nest by nest, amidst the mighty pillars and broken porticos of temples and forgotten edifices. The date tree elevated to a great height its broad leafy roof, as if it wished to form a shelter from the sun. The grey pyramids stood with their outlines sharply defined in the clear air towards the desert, where the ostrich knew he could use his legs; and the lion sat with his large grave eyes, and gazed on the marble sphinxes that lay half imbedded in the sand. The waters of the Nile had receded, and a great part of the bed of the river was swarming with frogs; and that, to the stork family, was the pleasantest sight in the country where they had arrived. The young ones were astonished at all they saw.

"Such are the sights here, and thus it always is in our warm country," said the stork-mother good-humouredly.

"Is there yet more to be seen?" they asked. "Shall we go much further into the country?"

"There is nothing more worth seeing," replied the stork-mother. "Beyond this luxuriant neighbourhood there is nothing but wild forests, where the trees grow close to each other, and are still more closely entangled by prickly creeping plants, weaving such a wall of verdure, that only the elephant, with his strong clumsy feet, can there tread his way. The snakes are too large for us there, and the lizards too lively. If ye would go to the desert, ye will meet with nothing but sand; it will fill your eyes, it will come in gusts, and cover your feathers. No, it is best here. Here are frogs and grasshoppers. I shall remain here, and so shall you."

And they remained. The old ones sat in their nest upon the graceful minaret; they reposed themselves, and yet they had enough to do to smooth their wings and rub their beaks on their red stockings; and they stretched out their necks, saluted gravely, and lifted up their heads with their high foreheads and fine soft feathers, and their brown eyes looked so wise.

The female young ones strutted about proudly among the juicy reeds, stole sly glances at the other young storks, made acquaintances, and slaughtered a frog at every third step, or went lounging about with little snakes in their bills, which they fancied looked well, and which they knew would taste well.

The male young ones got into quarrels; struck each other with their wings; pecked at each other with their beaks, even until blood flowed. Then they all thought of engaging themselves — the male and the female young ones. It was for that they lived, and they built nests, and got again into new quarrels; for in these warm countries every one is so hot-headed. Nevertheless they were very happy, and this was a great joy to the old storks. Every day there was warm sunshine — every day plenty to eat. They had nothing to think of except pleasure. But yonder, within the splendid palace of their Egyptian host, as they called him, there was but little pleasure to be found.

The wealthy, mighty chief lay upon his couch, stiffened in all his limbs — stretched out like a mummy in the centre of the grand saloon with the many-coloured painted walls: it was as if he were lying in a tulip. Kinsmen and servants stood around him. Dead he was not, yet it could hardly be said that he lived. The healing bog-flower from the far-away lands in the north — that which she was to have sought and plucked for him — she who loved him best — would never now be brought. His beautiful young daughter, who in the magic garb of a swan had flown over sea and land away to the distant north, would never more return. "She is dead and gone," had the two swan ladies, her companions, declared on their return home. They had concocted a tale, and they told it as follows : —

"We had flown all three high up in the air when a sportsman saw us, and shot at us with his arrow. It struck our young friend; and, slowly singing her farewell song, she sank like a dying swan down into the midst of the lake in the wood. There, on its banks, under a fragrant weeping birch tree, we buried her. But we took a just revenge: we bound fire under the wings of the swallow that built under the sportman's thatched roof. It kindled — his house was soon in flames — he was burned within it — and the flames shone as far over the sea as to the drooping birch, where she is now earth within the earth. Alas! never will she return to the land of Egypt."

And they both wept bitterly; and the old stork-father, when he heard it, rubbed his bill until it was quite sore.

"Lies and deceit!" he cried. "I should like, above all things, to run my beak into their breasts."

"And break it off," said the stork-mother; "you would look remarkably well then. Think first of yourself, and the interests of your own family; everything else is of little consequence."

"I will, however, place myself upon the edge of the open cupola to-morrow, when all the learned and the wise are to assemble to take the case of the sick man into consideration: perhaps they may then arrive a little nearer to the truth."

And the learned and the wise met together, and talked much, deeply, and profoundly of which the stork could make nothing at all; and, sooth to say, there was no result obtained from all this talking, either for the invalid or for his daughter in "the wild morass;" yet, nevertheless, it was all very well to listen to — one *must* listen to a great deal in this world.

But now it were best, perhaps, for us to hear what had happened formerly. We shall then be better acquainted with the story — at least, we shall know as much as the stork-father did.

"Love bestows life; the highest love bestows the highest life; it is only through love that his life can be saved," was what had been said; and it was amazingly wisely and well said, the learned declared.

"It is a beautiful thought," said the stork-father.

"I don't quite comprehend it," said the stork-mother, "but that is not my fault — it is the fault of the thought;

E

though it is all one to me, for I have other things to think upon."

And then the learned talked of love between this and that — that there was a difference. Love such as lovers felt, and that between parents and children ; between light and plants; how the sunbeams kissed the ground, and how thereby the seeds sprouted forth — it was all so diffusely and learnedly expounded, that it was impossible for the stork-father to follow the discourse, much less to repeat it. It made him very thoughtful, however ; he half closed his eyes, and actually stood on one leg the whole of the next day, reflecting on what he had heard. So much learning was difficult for him to digest.

But this much the stork-father understood. He had heard both common people and great people speak as if they really felt it, that it was a great misfortune to many thousands, and to the country in general, that the king lay so ill, and that nothing could be done to bring about his recovery. It would be a joy and a blessing to all if he could but be restored to health.

"But where grew the health-giving flower that might cure him?" Everybody asked that question. Scientific writings were searched, the glittering stars were consulted, the wind and the weather. Every traveller that could be found was appealed to, until at length the learned and the wise, as before stated, pitched upon this : "Love bestows life — life to a father." And though this dictum was really not understood by themselves, they adopted it, and wrote it out as a prescription. "Love bestows life" — well and good. But how was this to be applied? Here they were at a stand. At length, however, they agreed that the princess must be the means of procuring the necessary help, as she loved her father with all her heart and soul. They also agreed on a mode of proceeding. It is more than a year and a day since then. They settled that, when the new moon had just disappeared, she was to betake herself by night to the marble sphinx in the desert, to remove the sand from the entrance with her foot, and then to follow one of the long passages which led to the centre of the great pyramids, where one of the most mighty monarchs of ancient times, surrounded by splendour and magnificence, lay in his mummy-coffin. There she was to lean her head over the corpse, and then it would be revealed to her where life and health for her father were to be found.

All this she had performed, and in a dream had been instructed that from the deep morass high up in the Danish land — the place was minutely described to her — she might bring home a certain lotus flower, which beneath the water would touch her breast, that would cure him.

And therefore she had flown, in the magical disguise of a swan, from Egypt up to "the wild morass." All this was well known to the stork-father and the stork-mother: and now, though rather late, we also know it. We know that the mud-king dragged her down with him, and that, as far as regarded her home, she was dead and gone; only the wisest of them all said, like the stork-mother, "She can take care of herself;" and, knowing no better, they waited to see what would turn up.

"I think I shall steal their swan garbs from the two wicked princesses," said the stork-father; "then they

will not be able to go to 'the wild morass' and do mischief. I shall leave the swan disguises themselves up
yonder till there is some use for them."

"Where could you keep them?" asked the old female
stork.

"In our nest near 'the wild morass,'" he replied.
"I and our eldest young ones can carry them; and if we
find them too troublesome, there are plenty of places on
the way where we can hide them until our next flight.
One swan's dress would be enough for her, to be sure;
but two are better. It is a good thing to have abundant
means of travelling at command in a country so far
north."

"You will get no thanks for what you propose doing,"
said the stork-mother; "but you are the master, and
must please yourself. I have nothing to say except at
hatching-time."

At the Viking's castle near "the wild morass," whither the storks were flying in the spring, the little girl had
received her name. She was called Helga; but this
name was too soft for one with such dispositions as that
lovely creature had. She grew fast month by month;
and in a few years, even while the storks were making
their habitual journeys in autumn towards the Nile, in
spring towards "the wild morass," the little child had
grown up into a big girl, and before any one could have
thought it, she was in her sixteenth year, and a most
beautiful young lady — charming in appearance, but hard
and fierce in temper — the most savage of the savage in
that gloomy, cruel time.

It was a pleasure to her to sprinkle with her white hands the recking blood of the horse slaughtered for an offering. She would bite, in her barbarous sport, the neck of the black-cock which was to be slaughtered by the sacrificial priest; and to her foster-father she said in positive earnestness, —

"If your enemy were to come and cast ropes over the beams that support the roof, and drag them down upon your chamber whilst you were sleeping, I would not awaken you if I could — I would not hear it — the blood would tingle as it does now in that ear on which, years ago, you dared to give me a blow. I remember it well."

But the Viking did not believe she spoke seriously. Like every one else, he was fascinated by her extreme beauty, and never troubled himself to observe if the mind of little Helga were in unison with her looks. She would sit on horseback without a saddle, as if grown fast to the animal, and go at full gallop; nor would she spring off, even if her horse and other ill-natured ones were biting each other. Entirely dressed as she was, she would cast herself from the bank into the strong current of the fiord, and swim out to meet the Viking when his boat was approaching the land. Of her thick, splendid hair she had cut off the longest lock, and plaited for herself a string to her bow.

"Self-made is well made," she said.

The Viking's wife, according to the manners and customs of the age in which she lived, was strong in mind, and decided in purpose; but with her daughter she was like a soft, timid woman. She was well aware that the dreadful child was under the influence of sorcery.

And Helga apparently took a malicious pleasure in frightening her mother. Often when the latter was standing on the balcony, or walking in the courtyard, Helga would place herself on the side of the well, throw her arms up in the air, and then let herself fall headlong into the narrow, deep hole, where, with her frog nature, she would duck and raise herself up again, and then crawl up as if she had been a cat, and run dripping of water into the grand saloon, so that the green rushes which were strewed over the floor partook of the wet stream.

There was but one restraint upon little Helga — that was the *evening twilight*. In it she became quiet and thoughtful — would allow herself to be called and guided; then too she would seem to feel some affection for her mother; and when the sun sank, and the outer and inward change took place, she would sit still and sorrowful, shrivelled up into the form of a frog, though the head was now much larger than that little animal's, and therefore she was uglier than ever: she looked like a miserable dwarf, with a frog's head and webbed fingers. There was something very sad in her eyes; voice she had none except a kind of croak like a child sobbing in its dreams. Then would the Viking's wife take her in her lap; she would forget the ugly form, and look only at the melancholy eyes; and more than once she exclaimed, —

" I could almost wish that thou wert always my dumb fairy-child, for thou art more fearful to look at when thy form resumes its beauty."

And she wrote Runic rhymes against enchantment and infirmity, and threw them over the poor creature; but there was no change for the better.

"One could hardly believe that she was once so small as to lie in the calyx of a water-lily," said the stork-father. She is now quite a woman, and the image of her Egyptian mother. Her, alas! we have never seen again. She did not take good care of herself, as thou didst expect and the learned people predicted. Year after year I have flown backwards and forwards over 'the wild morass,' but never have I seen a sign of her. Yes, I can assure thee, during the years we have been coming up here, when I have arrived some days before thee, that I might mend the nest and set everything in order in it, I have for a whole night flown, as if I had been an owl or a bat, continually over the open water, but to no purpose. We have had no use either for the two swan disguises which I and the young ones dragged all the way up here from the banks of the Nile. It was hard enough work, and it took us three journeys to bring them up. They have now lain here for years at the bottom of our nest; and should a fire by any chance break out, and the Viking's house be burned down, they would be lost."

"And our good nest would be lost," said the old female stork; "but thou thinkest less of that than of these feather things and thy bog princess. Thou hadst better go down to her at once, and remain in the mire. Thou art a hard-hearted father to thine own: *that* I have said since I laid my first eggs. What if I or one of our young ones should get an arrow under our wings from that fierce crazy brat at the Viking's? She does not care what she does. This has been much longer our home than hers, she ought to recollect. We do not forget our duty; we pay our rent every year — a feather, an egg, and a young one — as we ought to do.   Dost thou think that when *she* is outside *I* can venture to go below, as in former days, or as I do in Egypt, where I am almost everybody's comrade, not to mention that I can there even peep into the pots and pans without any fear?   No; I sit up here and fret myself about her — the hussy! and I fret myself at thee too.   Thou shouldst have left her lying in the waterlily, and there would have been an end of her."

"Thy words are much harder than thy heart," said the stork-father.   "I know thee better than thou knowest thyself."

And then he made a hop, flapped his wings twice, stretched his legs out behind him, and away he flew, or rather sailed, without moving his wings, until he had got to some distance.   Then he brought his wings into play; the sun shone upon his white feathers; he stretched his head and his neck forward, and hastened on his way.

"He is, nevertheless, still the handsomest of them all," said his admiring mate; "but I will not tell him that."

Late that autumn the Viking returned home, bringing with him booty and prisoners.   Among these was a young Christian priest, one of the men who denounced the gods of the Northern mythology.   Often about this time was the new religion talked of in baronial halls and ladies' bowers — the religion that was spreading over all lands of the south, and which, with the holy Ansgarius,*

* Ansgarius was originally a monk from the monastery of New Corbie, in Saxony, to which several of the monks of Corbie in France had migrated in A. D. 822.   Its abbot, Paschasius Radbert, who died in 865, was, according to Cardinal Bellarmine, the first fully to propa-

had even reached as far as Hedeby. Even little Helga had heard of the pure religion of Christ, who, from love to mankind, had given himself as a sacrifice to save them; but with her it went in at one ear and out at the other, to use a common saying. The word *love* alone seemed to have made some impression upon her, when she shrunk into the miserable form of a frog in the closed-up chamber. But the Viking's wife had listened to, and felt herself wonderfully affected by, the rumour and the Saga about the Son of the one only true God.

The men, returning from their expedition, had told of the splendid temples of costly hewn stone raised to Him whose errand was love. A pair of heavy golden vessels, beautifully wrought out of pure gold, were brought home, and both had a charming, spicy perfume. They were the censers which the Christian priests swung before the altars, on which blood never flowed; but wine and the consecrated bread were changed into the blood of Him who had given himself for generations yet unborn.

---

gate the belief, now entertained in the Roman Catholic Church, of the corporeal presence of the Saviour in the sacrament. Ansgarius, who was very enthusiastic, accepted a mission to the north of Europe, and preached Christianity in Denmark and Sweden. Jutland was for some time the scene of his labours, and he made many converts there; also in Sleswig, where a Christian school for children was established, who, on leaving it, were sent to spread Christianity throughout the country. An archbishopic was founded by the then Emperor of Germany in conformity to a plan which had been traced, though not carried out, by Charlemagne; and this was bestowed upon Ansgarius. But the church he had built was burnt by some still heathen Danes, who, gathering a large fleet, invaded Hamburg, which they also reduced to ashes. The Emperor then constituted him Bishop of Bremen. — *Trans.*

4

To the deep, stone-walled cellars of the Viking's log-house was the young captive, the Christian priest, con-signed, fettered with cords round his feet and his hands. He was as beautiful as Baldur to look at, said the Vi-king's wife, and she was grieved at his fate; but young Helga wished that he should be ham-strung, and bound to the tails of wild oxen.

"Then I should let loose the dogs. Halloo! Then away over bogs and pools to the naked heath. Hah! that would be something pleasant to see — still pleasanter to follow him on the wild journey."

But the Viking would not hear of his being put to such a death. On the morrow, as a scoffer and denier of the high gods, he was to be offered up as a sacrifice to them upon the blood stone in the sacred grove. He was to be the first human sacrifice ever offered up there.

Young Helga prayed that she might be allowed to sprinkle with the blood of the captive the images of the gods and the assembled spectators. She sharpened her gleaming knife, and, as one of the large ferocious dogs, of which there were plenty in the courtyard, leaped over her feet, she stuck the knife into his side.

"That is to prove the blade," she exclaimed.

And the Viking's wife was shocked at the savage-tem-pered, evil-minded girl; and when night came, and the beauteous form and disposition of her daughter changed, she poured forth her sorrow to her in warm words, which came from the bottom of her heart.

The hideous frog with the ogre head stood before her, and fixed its brown sad eyes upon her, listened, and seemed to understand with a human being's intellect.

"Never, even to my husband, have I hinted at the double sufferings I have through you," said the Viking's wife. "There is more sorrow in my heart on your account than I could have believed. Great is a mother's love. But love never enters your mind. Your heart is like a lump of cold hard mud. From whence did you come to my house?"

Then the ugly shape trembled violently; it seemed as if these words touched an invisible tie between the body and the soul — large tears started to its eyes.

"Your time of trouble will come some day, depend on it," said the Viking's wife, "and dreadful will it also be for me. Better had it been had you been put out on the highway, and the chillness of the night had benumbed you until you slept in death;" and the Viking's wife wept salt tears, and went angry and distressed away, passing round behind the loose skin partition that hung over an upper beam to divide the chamber.

Alone in a corner sat the shrivelled frog. She was mute, but after a short interval she uttered a sort of half-suppressed sigh. It was as if in sorrow a new life had awoke in some nook of her heart. She took a step forward, listened, advanced again, and grasping with her awkward hands the heavy bar that was placed across the door, she removed it softly, and quietly drew away the pin that was stuck in over the latch. She then seized the lighted lamp that stood in the room beyond: it seemed as if a great resolution had given her strength. She made her way down to the dungeon, drew back the iron bolt that fastened the trap-door, and slid down to where the prisoner was lying. He was sleeping. She touched him with her cold, clammy hand; and when he awoke, and beheld the disgusting creature, he shuddered as if he had seen an evil apparition. She drew her knife, severed his bonds, and beckoned to him to follow her.

He named holy names, made the sign of the cross, and when the strange shape stood without moving, he exclaimed, in the words of the Bible, —

"'Blessed is he that considereth the poor: the Lord will deliver him in time of trouble.' Who art thou? How comes it that, under the exterior of such an animal, there is so much compassionate feeling?"

The frog beckoned to him, and led him, behind tapestry that concealed him, through private passages out to the stables, and pointed to a horse. He sprang on it, and she also jumped up; and, placing herself before him, she held by the animal's mane. The prisoner understood her movement; and at full gallop they rode, by a path he never could have found, away to the open heath.

He forgot her ugly form — he knew that the grace and mercy of God could be evinced even by means of hobgoblins — he put up earnest prayers, and sang holy hymns. She trembled. Was it the power of the prayers and hymns that affected her thus? or was it a cold shivering at the approach of morning, that was about to dawn? What was it that she felt? She raised herself up into the air, attempted to stop the horse, and was on the point of leaping down; but the Christian priest held her fast with all his might, and chanted a psalm, which he thought would have sufficient strength to overcome the influence of the witchcraft under which she was kept in the hide-

ous disguise of a frog   And the horse dashed more wildly
forward, the heavens became red, the first ray of the sun
burst forth through the morning sky, and with that clear
gush of light came the miraculous change — she was the
young beauty, with the cruel, demoniacal spirit.   The
astonished priest held the loveliest maiden in his arms he
had ever beheld; but he was horror-struck, and, springing from the horse, he stopped it, expecting to see it also
the victim of some fearful sorcery.   Young Helga sprang
at the same moment to the ground, her short childlike
dress reaching no lower than her knees.   Suddenly she
drew her sharp knife from her belt, and rushed furiously
upon him.

"Let me but reach thee — let me but reach thee, and
my knife shall find its way to thy heart.   Thou art pale
in thy terror, beardless slave!"

She closed with him; a severe struggle ensued, but it
seemed as if some invincible power bestowed strength
upon the Christian priest.   He held her fast; and the
old oak tree close by came to his assistance by binding
down her feet with its roots, which were half loosened
from the earth, her feet having slid under them.   There
was a fountain near, and he splashed the clear, fresh
water over her face and neck, commanding the unclean
spirit to pass out of her, and signed her according to the
Christian rites; but the baptismal water had no power
where the fountain of belief had not streamed upon the
heart.

Yet still he was the victor.   Yes, more than human
strength could have accomplished against the powers of
evil lay in his acts, which, as it were, overpowered her.

She suffered her arms to sink, and gazed with wondering
looks and blanched cheeks upon the man whom she
deemed some mighty wizard, strong in sorcery and the
black art. These were mystic Rhunes he had recited,
and magic characters he had traced in the air. Not for
the glancing axe or the well-sharpened knife, if he had
brandished these before her eyes, would they have
blinked, or would she have winced; but she winced now
when he made the sign of the cross upon her brow and
bosom, and she stood now like a tame bird, her head
bowed down upon her breast.

Then he spoke kindly to her of the work of mercy she
had performed towards him that night, when, in the ugly
disguise of a frog, she had come to him, had loosened his
bonds, and brought him forth to light and life. She also
was bound — bound even with stronger fetters than he
had been, he said; but she also should be set free, and
like him attain to light and life. He would take her to
Hedeby, to the holy Ansgarius. There, in the Christian
city, the witchcraft in which she was held would be exor-
cised; but not before him must she sit on horseback, even
if she wished it herself — he dared not place her there.

"Thou must sit behind me on the horse, not before me.
Thine enchanting beauty has a magic power bestowed by
the evil one. I fear it; and yet the victory shall be
mine through Christ."

He knelt down and prayed fervently. It seemed as if
the surrounding wood had been consecrated into a holy
temple; the birds began to sing, as if they belonged to
the new congregation; the wild thyme sent forth its fragrant scent, as if to take the place of incense; while the

priest proclaimed these Bible words: "To give light to
them that sit in darkness, and in the shadow of death; to
guide our feet into the way of peace."

And he spoke of everlasting life; and as he discoursed,
the horse which had carried them in their wild flight
stood still, and pulled at the large bramble berries, so
that the ripest ones fell on little Helga's hand, inviting her
to pluck them for herself.

She allowed herself patiently to be lifted upon the
horse, and she sat on its back like a somnambulist, who
was neither in a waking nor a sleeping state. The Christian priest tied two small green branches together in the
form of a cross, which he held high aloft; and thus they
rode through the forest, which became thicker and thicker,
and the path, if path it could be called, taking them farther into it. The blackthorn stood as if to bar their way,
and they had to ride round outside of it; the trickling
streams swelled no longer into mere rivulets, but into
stagnant pools, and they had to ride round them; but as
the soft wind that played among the foliage of the trees
was refreshing and strengthening to the travellers, so the
mild words that were spoken in Christian charity and
truth served to lead the benighted one to light and life.

It is said that a constant dripping of water will make
a hollow in the hardest stone, and that the waves of the
sea will in time round the edges of the sharpest rocks.
The dew of grace which fell for little Helga softened the
hard, and smoothed the sharp, in her nature. True, it
was not discernible yet in her, nor was she aware of it
herself. What knows the seed in the ground of the
effect which the refreshing dew and the warm sunbeams
are to have in producing from it vegetation and flowers?

As a mother's song to her child, unmarked, makes an impression upon its infant mind, and it prattles after her several of the words without understanding them, but in time these words arrange themselves into order, and they become clearer, so in the case of Helga worked *that word* which is mighty to save.

They rode out of the forest, and crossed an open heath; then again they entered a pathless wood, where, towards evening, they encountered a band of robbers.

" Whence didst thou steal that beautiful wench?" they shouted, as they stopped the horse, and dragged its two riders down; for they were strong and robust men. The priest had no other weapon than the knife which he had taken from little Helga. With that he now stood on his defence. One of the robbers swung his ponderous axe, but the young Christian fortunately sprang aside in time to avoid the blow, which then fell on the unfortunate horse, and the sharp edge entered into its neck; blood streamed from the wound, and the poor animal fell to the ground. Helga, who had only at that moment awoke from her long deep trance, sprang forward, and cast herself over the gasping creature. The Christian priest placed himself before her as a shield and protection from the lawless men; but one of them struck him on the forehead with an iron hammer, so that it was dashed in, and the blood and brains gushed forth, while he fell down dead on the spot.

The robbers seized Helga by her white arms; but at that moment the sun went down, its last beam faded away, and she was transformed into a hideous-looking frog. The pale green mouth stretched itself over half the face, its arms became thin and slimy, and a broad hand, with webbed-like membranes, extended itself like a fan. Then the robbers withdrew their hold of her in terror and astonishment. She stood like the ugly animal among them, and, according to the nature of a frog, she began to hop about, and, jumping faster than usual, she soon escaped into the depths of the thicket. The robbers were then convinced that it was some evil artifice of the mischief-loving Loke, or else some secret magical deception; and in dismay they fled from the place.

The full moon had risen, and its silver light penetrated even the gloomy recesses of the forest, when from among the low thick brushwood, in the frog's hideous form, crept the young Helga. She stopped when she reached the bodies of the Christian priest and the slaughtered horse; she gazed on them with eyes that seemed full of tears, and the frog uttered a sound that somewhat resembled the sob of a child who was on the point of crying. She threw herself first over the one, then over the other; then took water up in her webbed hand, and poured it over them; but all was in vain — they were dead, and dead they would remain. She knew that. Wild beasts would soon come and devour their bodies. No, that must not be; therefore she determined to dig a grave in the ground for them, but she had nothing to dig it with except the branch of a tree and both her own hands. With these she worked away until her fingers bled. She found she made so little progress, that she feared the work would never be completed. Then she took water, and washed the dead man's face; covered it

4 *   F

with fresh green leaves; brought large boughs of the
trees, and laid them over him; sprinkled dead leaves
amongst the branches; fetched the largest stones she
could carry, and placed them over the bodies, and filled
up the openings with moss. When she had done all this
she thought that their tomb might be strong and safe ;
but during her long and arduous labour the night had
passed away. The sun arose, and young Helga stood
again in all her beauty, with bloody hands, and, for the
first time, with tears on her blooming cheeks.

During this change it seemed as if two natures were
wrestling within her; she trembled, looked around her as
if awakening from a painful dream, then seized upon the
slender branch of a tree near, and held fast by it as if
for support ; and in another moment she climbed like a
cat up to the top of the tree, and placed herself firmly
there. For a whole long day she sat there like a frightened squirrel in the deep loneliness of the forest, where
all is still and dead, people say. Dead! There flew by
butterflies chasing each other either in sport or in strife.
There were ant-hills near, each covered with hundreds of
little busy labourers, passing in swarms to and fro. In
the air danced innumerable gnats; crowds of buzzing
flies swept past; lady-birds, dragon-flies, and other winged
insects, floated hither and thither ; earth-worms crept
forth from the damp ground; moles crawled about; otherwise it was still — *dead*, as people say and think.

None remarked Helga, except the jays that flew
screeching to the top of the tree where she sat; they
hopped on the branches around her with impudent curiosity, but there was something in the glance of her eye

that speedily drove them away; they were none the wiser about her, nor, indeed, was she about herself. When the evening approached, and the sun began to sink, the transformation time rendered a change of position necessary. She slipped down the tree, and, as the last ray of the sun faded away, she was again the shrivelled frog, with the webbed-fingered hands; but her eyes beamed now with a charming expression, which they had not worn in the beautiful form; they were the mildest, sweetest girlish eyes that glanced from behind the mask of a frog — they bore witness to the deeply-thinking human mind, the deeply-feeling human heart; and these lovely eyes burst into tears — tears of unfeigned sorrow.·

Close to the lately raised grave lay the cross of green boughs that had been tied together — the last work of him who was now dead and gone. Helga took it up, and the thought presented itself to her that it would be well to place it amidst the stones, above him and the slaughtered horse. With the sad remembrances thus awakened, her tears flowed faster; and in the fulness of her heart she scratched the same sign in the earth round the grave — it would be a fence that would decorate it so well. And just as she was forming, with both of her hands, the figure of the cross, her magic disguise fell off like a torn glove; and when she had washed herself in the clear water of the fountain near, and in amazement looked at her delicate white hands, she made the sign of the cross between herself and the dead priest; then her lips moved, then her tongue was loosened; and that name which so often, during the ride through the forest, she had heard spoken and chanted, became audible from her mouth — she exclaimed, "JESUS CHRIST!"

When the frog's skin had fallen off she was again
the beautiful maiden; but her head drooped heavily,
her limbs seemed to need repose — she slept.

Her sleep was only a short one, however; she awoke
about midnight, and before her stood the dead horse
full of life; its eyes glittered, and light seemed to
proceed from the wound in its neck.   Close to it the
dead Christian priest showed himself — "more beautiful than Baldur," the Viking's wife would have said;
and yet he came as a flash of fire.

There was an earnestness in his large, mild eyes,
a searching, penetrating look — grave, almost stern —
that thrilled the young proselyte to the utmost depths
of her heart.   Helga trembled before him; and her
memory awoke as if with the power it would exercise
on the great day of doom.   All the kindness that had
been bestowed on her, every affectionate word that had
been said to her, came back to her mind with an impression deeper than they had ever before made.   She
understood that it was love that, during the days of
trial here, had supported her — those days of trial in
which the offspring of a being with a soul, and a form
of mud, had writhed and struggled.   She understood
that she had only followed the promptings of her own
disposition, and done nothing to help herself.   All had
been bestowed on her — all had been ordained for her.
She bowed herself in lowly humility and shame before
Him who must be able to read every thought of the
heart; and at that moment she felt as if a purifying
flame darted through her — a light from the Holy Spirit.

"Daughter of the dust!" said the Christian priest,

"from dust, from earth hast thou arisen — from earth shalt thou again arise! A ray from God's invisible sun shall stream on thee. No soul shall be lost. But far off is the time when life takes flight into eternity. I come from the land of the dead. Thou also shalt once pass through the dark valley into yon lofty realms of brightness, where grace and perfection dwell. I shall not guide thee now to Hedeby for Christian baptism. First must thou disperse the slimy surface over the deep morass, draw up the living root of thy life and thy cradle, and perform thy appointed task, ere thou darest to seek the holy rite."

And he lifted her up on the horse, and gave her a golden censer like those she had formerly seen at the Viking's castle; and strong was the perfume which issued from it. The open wound on the forehead of the murdered man shone like a diadem of brilliants. He took the cross from the grave, and raised it high above him; then away they went through the air, away over the rustling woods, away over the mountains where the giant heroes are buried, sitting on the slaughtered steed. Still onward the phantom forms pursued their way; and in the clear moonlight glittered the gold circlet round their brows, and the mantle fluttered in the breeze. The magic dragon, who was watching over his treasures, raised his head and gazed at them. The hill dwarfs peeped out from their mountain recesses and plough-furrows. There were swarms of them, with red, blue, and green lights, that looked like the numerous sparks in the ashes of newly-burned paper.

Away over forest and heath, over limpid streams

and stagnant pools, they hastened towards "the wild morass," and over it they flew in wide circles. The Christian priest held aloft the cross, which looked as dazzling as burnished gold, and as he did so he chanted the mass hymns. Little Helga sang with him as a child follows its mother's song. She swung the censer about as if before the altar, and there came a perfume so strong, so powerful in its effect, that it caused the reeds and sedges to blossom; every sprout shot up from the deep bottom — everything that had life raised itself up; and with the rest arose a mass of water-lilies, which looked like a carpet of embroidered flowers. Upon it lay a sleeping female, young and beautiful. Helga thought she beheld herself mirrored in the calm water; but it was her mother whom she saw — the mud-king's wife — the princess from the banks of the Nile.

The dead Christian priest prayed that the sleeper might be lifted upon the horse. At first the latter sank under the additional burden, as if its body were but a winding-sheet fluttering in the wind; but the sign of the cross gave strength to the airy phantom, and all three rode on it to the solid ground.

Then crowed the cock at the Viking's castle, and the apparitions seemed to disappear in a mist, which was wafted away by the wind; but the mother and daughter stood together.

"Is that myself I behold in the deep water?" exclaimed the mother.

"Is that myself I see on the shining surface?" said the daughter.

And they approached each other till form met form in a warm embrace, and wildly the mother's heart beat when she perceived the truth.

"My child! my heart's own flower! my lotus from the watery deep!"

And she encircled her daughter with her arm, and wept. Her tears caused a new sensation to Helga — they were the baptism of love for her.

"I came hither in the magic disguise of a swan, and I threw it off," said the mother. "I sank through the swaying mire deep into the mud of the morass, which like a wall closed around me; but soon I perceived that I was in a fresher stream — some power drew me deeper and still deeper down. I felt my eyelids heavy with sleep — I slumbered and I dreamed. I thought that I was again in the interior of the Egyptian pyramid, but before me still stood the heaving alder trunk that had so terrified me on the surface of the morass. I saw the cracks in the bark, and they changed their appearance, and became hieroglyphics. It was the mummy's coffin I was looking at; it burst open, and out issued from it the monarch of a thousand years ago — the mummy form, black as pitch, dark and shining as a wood-snail, or as that thick slimy mud. It was the mud-king, or the mummy of the pyramids; I knew not which. He threw his arms around me, and I felt as if I were dying. I only felt that I was alive again when I found something warm on my breast, and there a little bird was flapping with its wings, twittering and singing. It flew from my breast high up in the dark, heavy space; but a long green string bound it still to me. I heard and I comprehended its tones and its longing: 'Freedom! Sunshine! To the father!' Then I thought of my father in my distant home, that dear sunny land — my life, my affection — and I loosened the cord, and let it flutter away home to my father. Since that hour I have not dreamed. I have slept a long, dark, heavy sleep until now, when the strange sounds and perfume awoke me and set me free."

That green tie between the mother's heart and the bird's wings, where now did it flutter? what now had become of it? The stork alone had seen it. The cord was the green stem; the knot was the shining flower — the cradle for that child who now had grown up in beauty, and again rested near her mother's heart.

And as they stood there embracing each other the stork-father flew in circles round them, hastened back to his nest, took from it the magic feather disguises that had been hidden away for so many years, cast one down before each of them, and then joined them as they raised themselves from the ground like two white swans.

" Let us now have some chat," said the stork-father, " now we understand each other's language, even though one bird's beak is not exactly made after the pattern of another's. It is most fortunate that you came to-night; to-morrow we should all have been away — the mother, the young ones, and myself. We are off to the . south. Look at me! I am an old friend from the country where the Nile flows, and so is the mother, though there is more kindness in her heart than in her tongue. She always believed that the princess would make her escape. The young ones and I brought these swan garbs up here. Well, how glad I am, and how fortunate it is that I am here still! At dawn of day we shall take our departure — a large party of storks. We shall fly foremost, and if you will follow us you will not miss the way. The young ones and myself will have an eye to you."

"And the lotus flower I was to have brought," said the Egyptian princess; "it shall go within the swan disguise, by my side, and I shall have my heart's darling with me. Then homewards — homewards!"

Then Helga said that she could not leave the Danish land until she had once more seen her foster-mother, the Viking's excellent wife. To Helga's thoughts arose every pleasing recollection, every kind word, even every tear her adopted mother had shed on her account; and, at that moment, she felt that she almost loved that mother best.

"Yes, we must go to the Viking's castle," said the stork; "there my young ones and their mother await me. How they will stare! The mother does not speak much; but, though she is rather abrupt, she means well. I will presently make a little noise, that she may know we are coming."

And he clattered with his bill as he and the swans flew close to the Viking's castle.

Within it all were lying in deep sleep. The Viking's wife had retired late to rest; she lay in anxious thought about little Helga, who now for full three days and nights had disappeared along with the Christian priest: she had probably assisted him in his escape, for it was her horse that was missing from the stables. By what power had

all this been accomplished? The Viking's wife thought
upon the wondrous works she had heard had been performed by the immaculate Christ, and by those who
believed on him and followed him. Her changing
thoughts assumed the shapes of life in her dreams; she
fancied she was still awake, lost in deep reflection; she
imagined that a storm arose — that she heard the sea
roaring in the east and in the west, the waves dashing
from the Kattegat and the North Sea; the hideous serpents which encircled the earth in the depths of the
ocean struggling in deadly combat. It was the night of
the gods — RAGNAROK, as the heathens called the last
hour, when all should be changed, even the high gods
themselves. The reverberating horn sounded, and forth
over the rainbow * rode the gods, clad in steel, to fight
the final battle; before them flew the winged Valkyries,
and the rear was brought up by the shades of the dead
giant-warriors; the whole atmosphere was illuminated
around them by the Northern lights, but darkness con-
quered all — it was an awful hour!

And near the terrified Viking's wife sat upon the floor
little Helga in the ugly disguise of the frog; and she
shivered and worked her way up to her foster-mother,
who took her in her lap, and, disgusting as she was in
that form, lovingly caressed her. The air was filled with
the sounds of the clashing of swords, the blows of clubs,
the whizzing of arrows, like a violent hail-storm. The
time was come when heaven and earth should be destroyed, the stars should fall, and all be swallowed up

* The Bridge of Heaven in the fables of the Scandinavian mythology. — *Trans.*

below in Surtur's fire ; but a new earth and a new heaven she knew were to come; the corn was to wave where the sea now rolled over the golden sands; the unknown God at length reigned ; and to him ascended Baldur, the mild, the lovable, released from the kingdom of death. He came; the Viking's wife beheld him — she recognised his countenance : it was that of the captive Christian priest. " Immaculate Christ ! " she cried aloud ; and whilst uttering this holy name she impressed a kiss upon the ugly brow of the frog-child. Then fell the magic disguise, and Helga stood before her in all her radiant beauty, gentle as she had never looked before, and with speaking eyes. She kissed her foster-mother's hands, blessed her for all the care and kindness which she, in the days of distress and trial, had lavished upon her; thanked her for the thoughts with which she had inspired her mind — thanked her for mentioning *that name* which she now repeated, " Immaculate Christ ! " and then lifting herself up in the suddenly adopted shape of a graceful swan, little Helga spread her wings widely out with the rustling sound of a flock of birds of passage on the wing, and in another moment she was gone.

The Viking's wife awoke, and on the outside of her casement were to be heard the same rustling and flapping of wings. It was the time, she knew, when the storks generally took their departure; it was them she heard. She wished to see them once more before their journey to the south, and bid them farewell. She got up, went out on the balcony, and then she saw, on the roof of an adjoining outhouse, stork upon stork, while all around the place, above the highest trees, flew crowds of them, wheeling in large circles; but below, on the brink of the well, where little Helga had but so lately often sat, and frightened her with her wild actions, sat now two swans, looking up at her with expressive eyes; and she remembered her dream, which seemed to her almost a reality. She thought of Helga in the appearance of a swan; she thought of the Christian priest, and felt a strange gladness in her heart.

The swans fluttered their wings and bowed their necks, as if they were saluting her; and the Viking's wife opened her arms, as if she understood them, and smiled amidst her tears and manifold thoughts.

Then, with a clattering of bills and a noise of wings, the storks all turned towards the south to commence their long journey.

"We will not wait any longer for the swans," said the stork-mother. "If they choose to go with us, they must come at once; we cannot be lingering here till the plovers begin their flight. It is pleasant to travel as we do in a family party, not like the chaffinches and strutting cocks. Among their species the males fly by themselves, and the females by themselves: that, to say the least of it, is not at all seemly. What a miserable sound the stroke of the swans' wings has compared with ours!"

"Every one flies in his own way," said the stork-father. "Swans fly slantingly, cranes in triangles, and plovers in serpentine windings."

"Name not serpents or snakes when we are about to fly up yonder," said the stork-mother. "It will only make the young ones long for a sort of food which they can't get just now."

" Are these the high hills, beneath yonder, of which I have heard?" asked Helga, in the disguise of a swan.

" These are thunder-clouds driving under us," replied her mother.

" What are these white clouds that seem so stationary?" asked Helga.

" These are the mountains covered with everlasting snow that thou seest," said her mother; and they flew over the Alps towards the blue Mediterranean.

" There is Africa! there is Egypt!" cried in joyful accents, under her swan disguise, the daughter of the Nile, as high up in the air she descried, like a whitish-yellow, billow-shaped streak, her native soil.

The storks also saw it, and quickened their flight.

" I smell the mud of the Nile and the wet frogs," exclaimed the stork-mother. " It makes my mouth water. Yes, now ye shall have nice things to eat, and ye shall see the marabout, the ibis, and the crane: they are all related to our family, but are not nearly so handsome as we are. They think a great deal, however, of themselves, particularly the ibis: he has been spoiled by the Egyptians, who make a mummy of him, and stuff him with aromatic herbs. I would rather be stuffed with living frogs; and that is what ye would all like also, and what ye shall be. Better a good dinner when one is living, than to be made a grand show of when one is dead. That is what I think, and I know I am right."

" The storks have returned," was told in the splendid house on the banks of the Nile, where, within the open hall, upon soft cushions, covered with a leopard's skin,

the king lay, neither living nor dead, hoping for the lotus
flower from the deep morass of the North.   His kindred
and his attendants were standing around him.

And into the hall flew two magnificent white swans —
they had arrived with the storks.   They cast off the dazzling magic feather garbs, and there stood two beautiful
women, as like each other as two drops of water.   They
leaned over the pallid, faded old man; they threw back
their long hair; and, as little Helga bowed over her
grandfather, his cheeks flushed, his eyes sparkled, life
returned to his stiffened limbs.   The old man rose hale
and hearty; his daughter and his granddaughter pressed
him in their arms, as if in a glad morning salutation after
a long heavy dream.

And there was joy throughout the palace, and in the
storks' nest also; but *there* the joy was principally for
the good food, the swarms of nice frogs; and whilst the
learned noted down in haste, and very carelessly, the
history of the two princesses and of the lotus flower as
an important event, and a blessing to the royal house,
and to the country in general, the old storks related the
history in their own way to their own family; but not
until they had all eaten enough, else these would have
had other things to think of than listening to any story.

"Now thou wilt be somebody," whispered the stork-
mother; "it is only reasonable to expect that."

"Oh! what should *I* be?" said the stork-father.   "And
what have *I* done?   Nothing!"

"Thou hast done more than all the others put together.
Without thee and the young ones the two princesses would

never have seen Egypt again, or cured the old man. Thou wilt be nothing! Thou shouldst, at the very least, be appointed court doctor, and have a title bestowed on thee, which our young ones would inherit, and their little ones after them. Thou dost look already exactly like an Egyptian doctor in my eyes."

The learned and the wise lectured upon "the fundamental notion," as they called it, which pervaded the whole tissue of events. "Love bestows life." Then they expounded their meaning in this manner: —

"The warm sunbeam was the Egyptian princess; she descended to the mud-king, and from their meeting sprang a flower ——"

"I cannot exactly repeat the words," said the stork-father, who had been listening to the discussion from the roof, and was now telling in his nest what he had heard. "What they said was not easy of comprehension, but it was so exceedingly wise that they were immediately rewarded with rank and marks of distinction. Even the prince's head cook got a handsome present — that was, doubtless, for having prepared the repast."

"And what didst thou get?" asked the stork-mother. "They had no right to overlook the most important actor in the affair, and that was thyself. The learned only babbled about the matter. But so it is always."

Late at night, when the now happy household reposed in peaceful slumbers, there was one who was still awake; and that was not the stork-father, although he was standing upon his nest on one leg, and dozing like a sentry. No; little Helga was awake, leaning over the balcony, and gazing through the clear air at the large

blazing stars, larger and brighter than she had ever seen
them in the North, and yet the same. She was thinking
upon the Viking's wife near "the wild morass"—upon
her foster-mother's mild eyes—upon the tears she had
shed over the poor frog-child, who was now standing under the light of the glorious stars, on the banks of the
Nile, in the soft spring air. She thought of the love in
the heathen woman's breast—the love she had shown
towards an unfortunate being, who in human form was as
vicious as a wild beast, and in the form of a noxious animal was horrible to look upon or to touch. She gazed at
the glittering stars, and thought of the shining circle on
the brow of the dead priest, when they flew over the forest and the morass. Tones seemed again to sound on her
ears—words she had heard spoken when they rode together, and she sat like an evil spirit there—words about
the great source of love, the highest love, that which in-
cluded all races and all generations. Yes, what was not
bestowed, won, obtained? Helga's thoughts embraced by
day, by night, the whole of her good fortune; she stood
contemplating it like a child who turns precipitately from
the giver to the beautiful gifts; she passed on to the increasing happiness which might come, and would come.
Higher and higher rose her thoughts, till she so lost herself in the dreams of future bliss that she forgot the
Giver of all good. It was the superabundance of youthful spirits which caused her imagination to take so bold a
flight. Her eyes were flashing with her thoughts, when
suddenly a loud noise in the court beneath recalled her
to mundane objects. She saw there two enormous ostriches running angrily round in a narrow circle. She

had never before seen these large heavy birds, who looked as if their wings were clipped; and when she asked what had happened to them, she heard for the first time the Egyptian legend about the ostrich.

Its race had once been beautiful, its wings broad and strong. Then one evening the largest forest birds said to it, " Brother, shall we fly to-morrow, God willing, to the river, and drink?" And the ostrich answered, " Yes, I will." At dawn they flew away, first up towards the sun, higher and higher, the ostrich far before the others. It flew on in its pride up towards the light; it relied upon its own strength, not upon the Giver of that strength; it did not say, " God willing." Then the avenging angel drew aside the veil from the streaming flames, and in that moment the bird's wings were burnt, and he sank in wretchedness to the earth. Neither he nor his species were ever afterwards able to raise themselves up in the air. They fly timidly — hurry along in a narrow space; they are a warning to mankind in all our thoughts and all our enterprises to say, " God willing."

And Helga humbly bowed her head, looked at the ostriches rushing past, saw their surprise and their simple joy at the sight of their own large shadows on the white wall, and more serious thoughts took possession of her mind, adding to her present happiness — inspiring brighter hopes for the future. What was yet to happen? The best for her, " God willing."

In the early spring, when the storks were about to go north again, Helga took from her arm a golden bracelet, scratched her name upon it, beckoned to the stork-father,

5   G

hung the gold band round his neck, and bade him carry
it to the Viking's wife, who would thereby know that her
adopted daughter lived, was happy, and remembered her.

" It is heavy to carry," thought the stork, when it was
hung round his neck; "but gold and honor must not be
flung away upon the high road. The stork brings luck —
they must admit that up yonder."

" Thou layest gold, and I lay eggs," said the stork-
mother; "but thou layest only once, and I lay every
year. But neither of us gets any thanks, which is very
vexatious."

" One knows, however, that one has done one's duty,"
said the stork-father.

" But that can't be hung up to be seen and lauded;
and if it could be, fine words butter no parsnips."

So they flew away.

The little nightingale that sang upon the tamarind tree
would also soon be going north, up yonder near " the
wild morass." Helga had often heard it — she would
send a message by it; for, since she had flown in the
magical disguise of the swan, she had often spoken to the
storks and the swallows. The nightingale would therefore understand her, and she prayed it to fly to the beech
wood upon the Jutland peninsula, where the tomb of
stone and branches had been erected. She asked it to
beg all the little birds to protect the sacred spot, and frequently to sing over it.

And the nightingale flew away, and time flew also.

And the eagle stood upon a pyramid, and looked in the
autumn on a stately procession with richly-laden camels,

with armed and splendidly equipped men on snorting
Arabian horses shining white like silver, with red trembling nostrils, with long thick manes hanging down to
their slender legs. Rich guests — a royal Arabian prince,
handsome as a prince should be — approached the gor-
geous palace where the storks' nests stood empty. Those
who dwelt in these nests were away in the far North, but
they were soon to return; and they arrived on the very
day that was most marked by joy and festivities. It
was a wedding feast; and the beautiful Helga, clad in
silk and jewels, was the bride. The bridegroom was the
young prince from Arabia. They sat at the upper end
of the table, between her mother and grandfather.

But she looked not at the bridegroom's bronzed and
manly cheek, where the dark beard curled. She looked
not at his black eyes, so full of fire, that were fastened
upon her. She gazed outwards upon the bright twink-
ling stars that glittered in the heavens.

Then a loud rustling of strong wings was heard in the
air. The storks had come back; and the old pair, fatigued as they were after their journey, and much in
need of rest, flew immediately down to the rails of the
verandah, for they knew what festival was going on.
They had heard already at the frontiers that Helga had
had them painted upon the wall, introducing them into
her own history.

"It was a kind thought of hers," said the stork-father.

"It is very little," said the stork-mother. "She could
hardly have done less."

And when Helga saw them she rose, and went out into
the verandah to stroke their backs. The old couple

bowed their necks, and the youngest little ones felt themselves much honoured by being so well received.

And Helga looked up towards the shining stars, that glittered more and more brilliantly; and between them and her she beheld in the air a transparent form. It floated nearer to her. It was the dead Christian priest, who had also come to her bridal solemnity—come from the kingdom of heaven.

"The glory and beauty up yonder far exceed all that is known on earth," he said.

And Helga pleaded softly, earnestly, that but for one moment she might be allowed to ascend up thither, and to cast one single glance on those heavenly scenes.

Then he raised her amidst splendour and magnificence, and a stream of delicious music. It was not around her only that all seemed to be brightness and music, but the light seemed to stream in her soul, and the sweet tones to be echoed there. Words cannot describe what she felt.

"We must now return," he said; "thou wilt be missed."

"Only one more glance!" she entreated. "Only one short minute!"

"We must return to earth—the guests are all departing."

"But one more glance—the last!"

And Helga stood again in the verandah, but all the torches outside were extinguished; all the light in the bridal saloon was gone; the storks were gone; no guests were to be seen — no bridegroom. All had vanished in these three short minutes.

Then Helga felt anxious. She wandered through the

vast empty halls — there slept foreign soldiers. She opened the side door which led to her own chambers, and, as she fancied she was entering them, she found herself in the garden: it had not stood there. Red streaks crossed the skies; it was the dawn of day.

Only three minutes in heaven, and a whole night on earth had passed away.

Then she perceived the storks. She called to them, spoke their language, and the old stork turned his head towards her, listened, and drew near.

"Thou dost speak our language," said he. "What wouldst thou? Whence comest thou, thou foreign maiden?"

"It is I — it is Helga! Dost thou not know me? Three minutes ago we were talking together in the verandah."

"That is a mistake," said the stork. "Thou must have dreamt this."

"No, no," she said, and reminded him of the Viking's castle, "the wild morass," the journey thence.

Then the old stork winked with his eyes.

"That is a very old story; I have heard it from my great-great-grandmother's time. Yes, truly there was once in Egypt a princess from the Danish land; but she disappeared on the evening of her wedding, many hundred years ago, and was never seen again. Thou canst read that thyself upon the monument in the garden, upon which are sculptured both swans and storks, and above it stands one like thyself in the white marble."

And so it was. Helga saw, comprehended it all, and sank on her knees.

The sun burst forth in all its morning splendour, and as, in former days, with its first rays fell the frog disguise, and the lovely form became visible; so now, in the baptism of light, arose a form of celestial beauty, purer than the air, as if in a veil of radiance to the Father above. The body sank into dust, and where she had stood lay a faded lotus flower!

"Well, this is a new finale to the story," said the stork-father, "which I by no means expected; but I am quite satisfied with it."

"I wonder what the young ones will say to it?" replied the stork-mother.

"Ah! that, indeed, is of the most consequence," said the stork-father.

The Quickest Runners.

THERE was a large reward offered — indeed, there were two rewards offered, a larger and a lesser one — for the greatest speed, not in one race alone, but to such as had got on fastest throughout the year.

"I got the highest prize," said the hare. "One had a right to expect justice when one's own family and best friends were in the council; but that the snail should have got the second prize I consider as almost an insult to me."

"No," observed the wooden fence, which had been a witness to the distribution of the prizes; "you must take diligence and good will into consideration. That remark was made by several very estimable persons, and that was also my opinion. To be sure the snail took half a year to cross the threshold; but he broke his thigh-bone in the tremendous exertion which that was for him. He devoted himself entirely to this race; and, moreover, he ran with his house on his back. All these weighed in his favour, and so he obtained the second prize."

"I think my claims might also have been taken into consideration," said the swallow. "More speedy than I,

in flight and motion, I believe no one has shown himself.
And where have I not been?   Far, far away!"

"And that is just your misfortune," said the wooden
fence. "You gad about too much. You are always on
the wing, ready to start out of the country when it begins to freeze. You have no love for your fatherland.
You cannot claim any consideration in it."

"But if I were to sleep all the winter through on the
moor," inquired the swallow — "sleep my whole time
away — should I be thus entitled to be taken into consideration?"

"Obtain an affidavit from the old woman of the moor
that you did sleep half the year in your fatherland, then
your claims will be taken into consideration."

"I deserved the first prize instead of the second," said
the snail. "I know very well that the hare only ran
from cowardice, whenever he thought there was danger
near. I, on the contrary, made the trial the business of
my life, and I have become a cripple in consequence of
my exertions. If any one had a right to the first prize
it was I; but I make no fuss; I scorn to do so."

"I can declare upon my honour that each prize, at
least as far as my voice in the matter went, was accorded
with strict justice," said the old sign-post in the wood,
who had been one of the arbitrators. "I always act
with due reflection, and according to order. Seven times
before have I had the honour to be engaged in the distribution of the prizes, but never until to-day have I had
my own way carried out. My plan has always hitherto
been thwarted — that was, to give the first prize to one
of the first letters in the alphabet, and the second prize

to one of the last letters. If you will be so good as to grant me your attention, I will explain it to you. The eighth letter in the alphabet from *A* is *H*— that stands for *Hare*, and therefore I awarded the greatest prize to the Hare; and the eighth letter from the end is *S*, therefore the *Snail* obtained the second prize. Next time the *I* will carry off the first prize, and *R* the second. A due attention to order and rotation should prevail in all rewards and appointments. Everything should go according to rule. *Rule* must precede merit."

"I should certainly have voted for myself, had I not been among the judges," said the mule. "People must take into account not only how quickly one goes, but what other circumstances are in question; as, for instance, how much one carries. But I would not this time have thought about that, neither about the hare's wisdom in his flight — his tact in springing suddenly to one side, to put his pursuers on the wrong scent, away from his place of concealment. No; there is one thing many people think much of, and which ought never to be disregarded. It is called THE BEAUTIFUL. I saw that in the hare's charming well-grown ears; it is quite a pleasure to see how long they are. I fancied that I beheld myself when I was little, and so I voted for him.

"Hush!" said the fly. "As for me, I will not speak; I will only say one word. I know right well that I have outrun more than one hare. The other day I broke the hind legs of one of the young ones. I was sitting on the locomotive before the train: I often do that. One sees so well there one's own speed. A young hare ran for a long time in front of the engine: he had no idea that I

5*

was there. At length he was just going to turn off the line, when the locomotive went over his hind legs and broke them, for I was sitting on it. The hare remained lying there, but I drove on. This was surely getting before him; but I do not care for the prize."

"It appears to me," thought the wild rose, but she did not say it — it is not her nature to express her ideas openly, though it might have been well had she done so — "it appears to me that the sunbeam should have had the first prize of honour, and the second also. It passes in a moment the immeasurable space from the sun down to us, and comes with such power that all nature is awakened by it. It has such beauty, that all we roses redden and become fragrant under it. The high presiding authorities do not seem to have noticed *it* at all. Were I the sunbeam, I would give each of them a sunstroke, that I would; but it would only make them crazy, and they will very likely be that without it. I shall say nothing," thought the wild rose. "There is peace in the wood; it is delightful to blossom, to shed refreshing perfume around, to live amidst the songs of birds and the rustling of trees; but the sun's rays will outlive us all."

"What is the first prize?" asked the earth-worm, who had overslept himself, and only now joined them.

"It gives free entrance to the kitchen garden," said the mule. "I proposed the prize, as a clear-sighted and judicious member of the meeting, with a view to the hare's advantage. I was resolved he should have it, and he is now provided for. The snail has permission to sit on the stone fence, and to enjoy the moss and the sunshine; and, moreover, he is appointed to be one of the chief judges of the next race. It is well to have one who is practically acquainted with the business in hand —on a committee, as human beings call it. I must say I expect great things from the future — we have made so good a beginning."

The Bell's Hollow.

"DING-DONG! ding-dong!" sounded from the buried bell in Odensee river. What sort of a river is that? Every child in the town of Odensee knows it. It flows round the foot of the gardens, from the locks to the water-mill, away under the wooden bridges. In the river grow yellow water-lilies, brown feather-like reeds, and the soft velvet-like bulrushes, so high and so large. Old, split willow trees, bent and twisted, hang far over the water by the side of the monks' meadows and the bleaching greens; but a little above is garden after garden — the one very different from the other; some with beautiful flowers and arbours, clean and in prim array, like dolls' villages; some only filled with cabbages; while in others there are no attempts at a garden to be seen at all, only great elder trees stretching themselves out, and hanging over the running water, which here and there is deeper than an oar can fathom.

Opposite to the nunnery is the deepest part. It is called "The Bell's Hollow," and there dwells the merman. He sleeps by day when the sun shines through the water, but comes forth on the clear starry nights, and by moonlight. He is very old. Grandmothers have heard of him from their grandmothers. They said he lived a lonely life, and had scarcely any one to speak to except the large old church bell. Once upon a time it hung up in the steeple of the church; but now there is no trace either of the steeple or the church, which was then called Saint Albani.

"Ding-dong! ding-dong!" rang the bell while it stood in the steeple; and one evening when the sun was setting, and the bell was in full motion, it broke loose, and flew through the air, its shining metal glowing in the red sunbeams. "Ding-dong! ding-dong! now I am going to rest," sang the bell; and it flew out to Odensee river, where it was deepest, and therefore that spot is now called "The Bell's Hollow." But it found neither sleep nor rest there. Down at the merman's it still rings; so that at times it is heard above, through the water, and many people say that its tones foretell a death; but there is no truth in that, for it rings to amuse the merman, who is now no longer alone.

And what does the bell relate? It was so very old, it was there before our grandmothers' grandmothers were born, and yet it was a child compared with the merman, who is an old, quiet, strange-looking person, with eel-skin leggings, a scaly tunic adorned with yellow water-lilies, a wreath of sedges in his hair, and weeds in his beard. It must be confessed he was not very handsome to look at.

It would take a year and a day to repeat all that the bell said, for it told the same old stories over and over again very minutely, making them sometimes

longer, sometimes shorter, according to its mood. It
told of the olden days — the rigorous, dark times.

To the tower upon St. Albani Church, where the bell
hung, ascended a monk. He was both young and handsome, but had an air of deep melancholy. He looked
through an aperture out over the Odensee river. Its
bed then was broad, and the monks' meadows were a
lake. He gazed over them, and over the green mound
called " The Nun's Hill," beyond which the cloister lay,
where the light shone from a nun's cell. He had known
her well, and he remembered the past, and his heart beat
wildly at the recollection.

" Ding-dong! ding-dong!"   This was one of the bell's
stories : —

" There came up to the tower one day an idiot servant
of the bishop ; and when I, the bell, who am cast in hard
and heavy metal, swung about and pealed, I could have
broken his head, for he seated himself immediately under
me, and began to play with two sticks, exactly as if it
had been a stringed instrument, and he sang to it thus :
' Now I may venture to sing aloud what elsewhere I dare
not whisper — sing of all that is kept hidden behind locks
and bolts. Yonder it is cold and damp. The rats eat
the living bodies. No one knows of it ; no one hears of
it — not even now, when the bell is pouring forth its loud-
est peal — ding-dong! ding-dong!'

"There was a king : he was called Knud. He hum-
bled himself both before bishops and monks ; but as he
unjustly oppressed the people, and laid heavy taxes on
them, they armed themselves with all sorts of weapons,
and chased him away as if he had been a wild beast.

He sought shelter in the church, and had the doors and windows closed. The furious multitude surrounded the sacred edifice, as I heard related; the crows and the ravens, and the jackdaws to boot, became scared by the noise and the tumult; they flew up into the tower, and out again; they looked on the multitude below, they looked also in at the church windows, and shrieked out what they saw.

"King Knud knelt before the altar and prayed; his brothers Erik and Benedict stood guarding him with their drawn swords; but the king's servitor, the false Blaké, betrayed his lord. They knew outside where he could be reached. A stone was cast in through the window at him, and the king lay dead. There were shouts and cries among the angry crowd, and cries among the flocks of frightened birds; and I joined them too. I pealed forth, ' Ding-dong! ding-dong!'

"The church bell hangs high, sees far around, receives visits from birds, and understands their language. To it whispers the wind through the wickets and apertures, and through every little chink; and the wind knows everything. He hears it from the air, for *it* encompasses all living things; it even enters into the lungs of human beings — it hears every word and every sigh. The air knows all, the wind repeats all, and the bell understands their speech, and rings it forth to the whole world — ' Ding-dong! ding-dong!'

"But all this was too much for me to hear and to know. I had not strength enough to ring it all out. I became so wearied, so heavy, that the beam from which I hung broke, and I flew through the luminous air down to where the river is deepest, where the merman dwells alone in solitude; and here I am, year after year, relating to him what I have seen and what I have heard. 'Ding-dong! ding-dong!'"

Thus rang the chimes from "The Bell's Hollow" in the Odensee river, as my grandmother declares.

But our schoolmaster says there is no bell ringing down there, for it could not be; and there is no merman down there, for there are no mermen; and, when all the church bells are ringing loudly, he says that it is not the bells, but the air that makes the sound. My grandmother told me that the bell also said this; so, since the schoolmaster and the bell agree in this, no doubt it is true.

The air knows everything. It is round us, it is in us; it speaks of our thoughts and our actions; and it proclaims them farther than did the bell now down in the Hollow in Odensee river, where the merman dwells — it proclaims all out into the great vault of heaven, far, far away, even into eternity, up to where the glorious bells of paradise peal in tones unknown to mortal ears.

Soup made of a Sausage-stick.

I.

"WE had a capital dinner yesterday," said an aged female mouse to one who had not been at the feast. " I sat only twenty-one from the old King of the Mice: that was not being badly placed. Shall I tell you what we had for dinner? It was all very well arranged. We had mouldy bread, the skin of bacon, tallow candles, and sausages. Twice we returned to the charge : it was as good as if we had had two dinners. There was nothing but good-humour and a pleasant chit-chat, as in an agreeable family circle. Not a mite was left except the sausage-stick. The conversation happened to fall upon the possibility of making soup of a sausage-stick. All said they had heard of it, but no one had ever tasted that soup, or knew how to prepare it. A health was proposed to the inventor, who, it was remarked, deserved to be superintendent of the poor. Was not that witty? And the old King of the Mice arose and declared that the one among the young mice who could prepare the soup in question most palatably should be his queen, and he would grant them a year and a day for the trial."

" Well, that was not a bad idea," said the other mouse.
" But how is the soup made? "

" Ay, how is it made? That was what they were all
asking, the young and the old. Every one was willing
enough to become the queen, but they were all loath to
take the trouble of going out into the world to acquire
the prescribed qualification; yet it was absolutely necessary to do so. But it does not suit every one to leave
her family and her snug old mouse-hole. One cannot be
going out every day after cheese parings, and sniffing the
rind of bacon. No: such pursuits, too often indulged in,
would perchance put them in the way of being eaten alive
by a cat."

These apprehensions were quite terrible enough to
scare most of the mice from going forth upon the search
of knowledge. Only four presented themselves for the
undertaking. They were young and active, but very
poor. They would have gone to the four corners of the
earth, if only good fortune might attend their enterprise.
Each of them took with her a sausage-stick to remind
her what she was travelling for. It was to be her walking staff.

On the 1st of May they set out, and on the 1st of May,
a year after, they returned; but only three of them.
The fourth did not report herself, and sent no tidings of
herself; and yet it was the day fixed for the royal de-
cision.

" There shall be no sadness or no drawback to our
pleasure," said the King of the Mice, as he gave orders
that every mouse within several miles round should be
invited. They were to assemble in the kitchen. The

three travelled mice were drawn up in a row alone. In the place of the fourth, who was absent, was deposited a sausage-stick covered with black crape. No one ventured to utter a word until the three had made their statements, and the king had determined what more was to be said.

We have now to hear all this.

II.

WHAT THE FIRST LITTLE MOUSE HAD SEEN AND LEARNT ON HER JOURNEY.

"When I first went forth into the wide world," said the little mouse, "I thought, as so many of my age do, that I had swallowed all the wisdom of the earth; but that was not the case — it required a year and a day for that to come to pass. I went at once to sea, on board a ship which was bound for the north. I had heard that cooks at sea were pretty well acquainted with their business; but there is little to do when one has plenty of sides of bacon, barrels of salt meat, and musty meal at hand. One lives delicately on these nice things; but one learns nothing like making soup of a sausage-stick. We sailed for many days and nights, and a stormy and wet time we had of it. When we reached our destination I left the vessel: this was far away up in the north.

"One has a strange feeling on leaving one's own mouse-hole at home, being carried away in a ship, which becomes a home for the time, and suddenly finding one's self, at the distance of more than a hundred miles, stand-

ing alone in a foreign land. I saw myself amidst a large
tangled wood full of pine and birch trees. Their scent
was so strong! It is not at all to my taste; but the perfume from the wild plants was so spicy that I was quite
charmed, and thought of the sausage and the seasoning
for the soup. There were lakes amidst the forest, the
water was beautifully clear close at hand, but looking in
the distance as black as ink. There were white swans
upon the lake. I mistook them at first for foam, they lay
so still; but when I saw them fly I recognised them.
They, however, belong to the race of geese. No one can
deny his kindred. I like mine, and I hastened to seek
the field mice, who, truth to tell, know very little except
what concerns their food; and it was just that on account
of which I had travelled to a foreign country. That any
one should think of making soup out of a sausage-stick
seemed to them so extraordinary an idea, that it was
speedily circulated through the whole wood; but that the
problem should be solved they considered an impossibil-
ity. Little did I think then that the very same night I
should be initiated into the process.

"It was midsummer; therefore it was that the woods
scented so strongly, they said; therefore were the plants
so aromatic in their perfume, the lake so clear, and yet
so dark with the white swans upon them. On the bor-
ders of the forest, amidst three or four houses, was erected
a pole as high as a mainmast, and around it hung wreaths
and ribbons. This was the Maypole. Girls and young
men danced round it, and sang to the accompaniment
of the fiddler's violin. All went on merrily till after the
sun had set, and the moon had risen, but I took no part

in the festivity; for what had a little mouse to do with a forest ball? I sat down amidst the soft moss, and held fast my sausage-stick. The moon shone brightly on a place where there was a solitary tree surrounded by moss so fine — yes, I venture to say as fine as the Mice-King's skin — but it had a green tint, and its colour was very soothing to the eye. All at once I saw approaching a set of the most beautiful little people, so little that they would only have reached to my knee; they looked like men and women, but they were better proportioned. They called themselves Elves, and their garments were composed of the leaves of flowers, trimmed with the wings of gnats and flies — not at all ugly. They seemed as if they were searching for something — what I did not know; but when they came a little nearer to me their leader tapped my sausage-stick, and said, 'This is what we want; it is all ready, all prepared;' and he became more and more joyful as he gazed upon my walking-stick.

"'You may borrow it, but not keep it,' said I.

"'Not keep it!' they all exclaimed together, as they seized my sausage-stick, and, dancing away to the green mossy spot, placed the sausage-stick there in the centre of it. They determined also on having a Maypole; and the stick they had just captured seeming quite suited to their purpose, it was soon ornamented.

"Small spiders spun gold threads around it — hung up waving veils and flags so finely worked, shining so snow-white under the moonbeams, that my eyes were quite dazzled. They took the colours from the wings of the butterflies, and sprinkled them on the white webs, till they seemed to be laden with flowers and diamonds. I did not know my own sausage-stick — it had become such a magnificent Maypole, that certainly had not its equal in the world. And now came tripping forwards the great mass of the elves, most of them very slightly clad; but what they did wear was of the finest materials. I looked on, of course, but in the background, for I was too big for them.

"Then what a game commenced! It was as if a thousand glass bells were ringing, the sound was so clear and full. I fancied the swans were singing, and I also thought I heard cuckoos and thrushes. At length it seemed as if the whole wood was filled with music. There were the sweet voices of children, the ringing of bells, and the songs of birds; and all these melodious sounds seemed to proceed from the elves' Maypole — an orchestra in itself — and that was my sausage-stick. I never would have believed that so much could have come from it; but much, of course, depended on what hands it fell into. I became very much agitated, and I wept, as a little mouse can weep, from sheer pleasure.

"The night was all too short; but, at this time of the year, the nights are not long up yonder. At the dawn of day there arose a fresh breeze; the surface of the lake became ruffled; all the delicately fine veils and flags disappeared in the air; the swinging kiosks of cobwebs, the suspension bridges and balustrades, or whatever they are called, which were constructed from leaf to leaf, vanished into nothing; six elves brought me my sausage-stick, and at the same time asked if I had any wish they could fulfil; whereupon I begged them to tell me how soup could be made from a sausage-stick.

"'What we can do,' said the foremost, laughing, 'you have just seen. You could scarcely have recognised your sausage-stick.'

"'You mean as you transformed it,' said I; and then I told them the cause of my journey, and what was expected at home from it. 'Of what use,' I asked, 'will it be to the King of the Mice and all our large community that I have seen this beautiful sight? I cannot shake the sausage-stick and say, You see here the stick — now comes the soup! That would be like a hoax.'

"Then the elf dipped its little finger into a blue violet, and said to me, —

"'Look! I spread a charm over your walking-stick, and when you return to the palace of the King of the Mice make it touch the king's warm breast, and violets will spring from every part of the staff, even in the coldest winter weather. See! you have now something worth taking home, and perhaps a little more.'"

But before the little mouse had finished repeating what the elf had said, she laid her staff against the king's breast, and sure enough there sprang forth from it the loveliest flowers. They yielded so strong a perfume that the king commanded that the mice who stood nearest the chimney should stick their tails in the fire, in order that the smell of the singed hair should overpower the odour from the flowers, which was very offensive.

"But what was 'the little more' you spoke of?" asked the King of the Mice.

"Oh!" said the little mouse, "it is what is called an *effect;*" and so she turned her sausage-stick. And behold, there were no more flowers visible! She held only the naked stick, and she moved it like a stick for beating time.

"The violets are for sight, smell, and touch, the elf told me; but there are still wanting hearing and taste."

She beat time, and there was music — not such, however, as sounded in the wood at the elfin fête; no, such as is heard at times in the kitchen. It came suddenly, like the wind whistling down the chimney. The pots and the pans boiled over, and the shovel thundered against the large brass kettle. It stopped as suddenly as it had commenced; and then was only to be heard the smothered song of the tea-kettle, which was so strange with its tones rising and falling, and the little pot and the large pot boiling, the one not troubling itself about the other, as if neither could think. Then the little mouse moved her time-stick faster and faster; the pots bubbled up and boiled over; the wind roared in the chimney; the commotion was so great that the little mouse herself got frightened, and dropped the stick.

"It was hard work to make that soup," cried the old king; "but where is the result — the dish?"

"That is all," said the little mouse, courtesying.

"All! Then let us hear what the next has to tell," said the king.

III.

WHAT THE SECOND MOUSE HAD TO RELATE.

"I WAS born in the palace library," said the second mouse. "I, and several members of my family there, have never had the good fortune to enter the dining-room, let alone the pantry. It was only when I first began my travels, and now again to-day, that I have even beheld a kitchen. We had often to endure hunger in the library, but we acquired much knowledge. The report of the reward offered by royalty for the discovery of the process by which soup could be made of a sausage-stick reached us even up there, and my grandmother thereupon looked for a manuscript which, though she could not read herself, she had heard read, wherein it was said, —

"'A poet can make soup out of a sausage-stick.'

"She asked me if I were a poet. I confessed I was not, to which she replied that I must go and try to become one. I begged to know what was to be done to acquire this art, for it appeared to me about as difficult to attain as to make the soup itself. But my grandmother had heard a good deal of reading, and she told me that the three things principally necessary were — good sense, imagination, and feeling. 'If thou canst go and furnish thyself with *these*, thou wilt be a poet; and there will be every chance of thy success in the matter of the sausage-stick.'

"So I set off to the westward, out into the wide world, to become a poet.

"*Good sense* I knew was the most important of all things, the two other qualities not being so highly esteemed. So I went first after good sense. Well, where did it dwell? 'Go to the ant; consider her ways, and be wise,' a great king of the Hebrews has said. I knew this from the library, and I never stopped until I reached

6

a large ant-hill; and there I settled myself to watch them.

"They are a very respectable tribe, the ants, and full of good sense; everything among them is as correctly done as a well-calculated sum in arithmetic. 'To labour and to lay eggs,' say they, 'is to live in the present, and to provide for the future;' and that they assuredly do. They divide themselves into the clean ants and the dirty ones. Rank is distinguished by a number. The queen ant is number one, and her will is their only law. She has swallowed all the wisdom, and it was of consequence to me to listen to her; but she said so much, and was so profoundly wise, that I could scarcely comprehend her.

"She said that their hill was the highest in the world; but close to the hill stood a tree that was higher, certainly much higher. She could not deny this, so she did not allude to it. One evening an ant had lost his way, and finding himself on the tree, he crept up the trunk, not as far as the top, but much higher than any ant had ever gone before; and when he descended, and found his way home at last, he imprudently told in the ant-hill of something much higher at a little distance from it. This was taken by one and all as an affront to the whole community, and the offending ant was condemned to have his mouth muzzled, as well as to perpetual solitude. But shortly after, another ant got as far as the tree, and made a similar journey and a similar discovery. He spoke of it, however, discreetly and mysteriously, and as he happened to be an ant of consideration — one of the clean — they believed him; and when he died they placed an egg-shell over him as a monument in honour of his extensive knowledge.

"I observed," said the little mouse, "that the ants
continually move with their eggs on their backs. One of
them dropped hers. She tried very hard to get it up
again, but could not succeed; then two others came and
helped her with all their might, until they had nearly lost
their own eggs, whereupon they let the attempt alone,
for one is nearest to one's self; and the queen ant remarked that both heart and good sense had been shown.
'These two qualities place us ants among reasonable
beings,' she said. 'Sense ought to be, and is, of the most
consequence; and I have the most of that;' and she
raised herself, in her self-satisfaction, on her hind leg.
I could not mistake her, and I swallowed her. 'Go to
the ant; consider her ways, and be wise.' I had now
the queen.

"I then went nearer to the above-mentioned large
tree: it was an oak. It had high branches, a majestic
crown of leaves, and was very old. I perceived that a
living creature resided in it — a female. She was called
a Dryad. She had been born with the tree and would
die with it. I had heard of this in the library; and now
I beheld one of the real trees, and a real oak-nymph.
She uttered a frightful shriek when she saw me near
her; for she was like all women, very much afraid of
mice. She, however, had more reason to be afraid of me
than others of her sex have, for I could have gnawed the
tree in two, and on it hung her life. I spoke to her kindly and cordially. This gave her courage, and she took
me in her slender hand; and when she understood what
had brought me out into the wide world, she promised
that I should, perhaps that very night, become possessed

of one of the two treasures of which I was in search. She told me that Imagination was her very particular friend; that he was as charming as the God of Love; and that he often, for many an hour, sought repose under the spreading foliage of the tree, which then sighed more musically over the two. He called her *his* dryad, she said, and the tree *his* tree. The mighty, gnarled, majestic oak was just to his taste, with its broad roots sunk deep into the earth, its trunk and its coronal rising so high in the free air, meeting the drifting snow, the cutting winds, and the bright sunshine, before they had reached the ground. All this she said, and she continued: 'The birds sing up yonder, and tell of foreign lands, and upon the only decayed branch the stork has built a nest; and it is a pleasure to hear of the country where the pyramids stand. All this Fancy can well depict, and very much more. I myself can describe life in the woods from the time that I was quite little, and this tree was so tiny that a nettle could have covered it, until now, when it is so strong and mighty. Sit down yonder under the woodruffs, and be on the look-out. When Fancy comes I shall find an opportunity of pinching his wing, and stealing a little feather from it. You shall take that, and no poet will ever have been better provided. Will that do?'

"And Imagination came; a feather was plucked from him, and I got it," said the little mouse. "I held it in the water till it became soft. It was still hard of digestion, but I managed to gnaw it all up. It is not at all easy to stuff one's self so as to be a poet — there is so much to be put in one. I had now got two of the in-

gredients — good sense and imagination ; and I knew by
their help that the third ingredient was to be found in
the library ; for a great man has said and written, that
there are romances which are useful in easing people
of a superfluity of tears, and which also act as a sort
of swamp to cast feelings into. I remembered some of
these books ; they had always looked very enticing to
me. They were so thumbed, so greasy, they must have
been very popular.

"I returned home to the library, ate almost as much
as a whole romance — that is to say, the soft part of it,
the pith — but the crust, the binding, I let alone. When
I had digested this, and another to boot, I perceived how
my inside was stirred up ; so I ate part of a third, and
then I considered myself a poet, and every one about
me said I was. I had headaches, of course, and all sorts
of aches. I thought over what story I could work up
about a sausage-stick, and there was no end of sticks and
pegs crowding my mind. The queen ant had had an un-
common intellect. I remembered the man who took a
white peg into his mouth, and both he and it became invisible. All my thoughts ran upon sticks. A poet can
write even upon these ; and I am a poet I trust, for I
have fagged hard to be one. I shall be able every day
in the week to amuse you with the story of a stick. This
is my soup."

" Let us hear the third, said the King of the mice.

" Pip, pip !" said a little mouse at the kitchen door.
It was the fourth of them, the one they thought dead.
She tripped in, and jumped upon the upper end of the
sausage-stick with the black crape. She had been jour-

neying day and night, travelling on the railroad by the
goods train, in which she took great pleasure, and yet
she had almost arrived too late; but she hurried forward
puffing and panting, and looking very much jaded. She
had lost her sausage-stick, but not her voice; for she began talking with the utmost velocity, as if every one was
dying to hear her, and no one could say anything to the
purpose but herself. How she did chatter! But she
had arrived so unexpectedly that no one had time to find
fault with her or her talking, so she went on. Now let
us listen.

IV.

WHAT THE FOURTH MOUSE — WHO SPOKE BEFORE THE THIRD ONE HAD SPOKEN — HAD TO RELATE.

"I went straight to the greatest city," she said. "I
do not remember its name. I do not recollect names
well. I came from the railway with confiscated goods to
the town council-hall, and there I ran to the jailer. He
spoke of his prisoners, especially of one of them, who
had uttered some very imprudent words; and when
these had been repeated, and written down and read,
'The whole,' said he, 'was only — soup of a sausage-
stick; but that soup may cost him dear.' I felt inter-
ested in the prisoner," continued the little mouse, "and
I watched for an opportunity to go in where he was.
There is always a mouse-hole behind locked doors. He
looked very pale, had a dark beard, and large shining
eyes. The lamp smoked; but the walls were accus-

tomed to this. They did not turn any blacker. The
prisoner was scratching on them both pictures and
verses; but I did not read the latter. I fancy he was
tired of being alone, for I was a welcome guest. He
enticed me with crumbs of bread, with his flute, and kind
words. He was so happy with me! I put confidence
in him, and we became friends. He shared with me
bread and water, and gave me cheese and sausages. I
lived luxuriously; but it was not alone the good cheer
that detained me. He allowed me to run upon his hand
and arm all the way up to his shoulder; he allowed me
to creep into his beard, and called me his little friend.
I became very dear to him, and our regard was mutual.
I forgot my errand out in the wide world; I forgot my
sausage-stick in a crevice in the floor; and there it still
lies. I wished to remain where I was; for, if I left him,
the poor prisoner would have nothing to care for in this
world. I remained; but he, alas! did not. He spoke
to me so sadly for the last time, gave me a double allow-
ance of bread and cheese parings, kissed his finger to
me, and then he was gone — gone, never to return. I
do not know his history. 'Soup of a sausage-stick!' said
the jailer, and I went to him; but I was wrong to trust
in him. He took me up, indeed, in his hand; but he put
me in a cage, a treadmill. That was hard work —
jumping and jumping without getting on a bit, and only
to be laughed at.

"The jailer's grandchild was a pretty little fellow, with
waving hair as yellow as gold, sparkling, joyous eyes, and
a laughing mouth.

"'Poor little mouse!' he exclaimed, peeping in at my

horrid cage, and at the same time drawing up the iron pin that closed it.

"I seized the opportunity, and sprang first to the window-ledge, and thence to the conduit-pipe. Free, free! that was all I could think of, and not the object of my journey.

"It became dark — it was almost night. I took up my lodgings in a tower, where dwelt a watchman and an owl. I could not trust either of them, and the owl least of the two. It resembles a cat, and has one great fault — that it eats mice. But one can be on one's guard, and that I assuredly would be. She was a respectable, extremely well-educated old owl. She knew more than the watchman, and almost as much as I myself did. The young owls made a great fuss about everything.

"'Don't make soup of a sausage-stick,' said she.

"This was the severest thing she could say to them, she was so very fond of her family. I felt so much inclined to place some reliance in her that I cried 'Pip!' from the crevice in which I was concealed. My confidence in her seemed to please her, and she assured me that I should be safe under her protection; that no animal would be permitted to injure me until winter, when she might herself fall upon me, as food would be scarce.

"She was very wise in all things. She proved to me that the watchman could not blow a blast without his horn, which hung loosely about him.

"He piques himself exceedingly upon his performances, and fancies he is the owl of the tower. The sound ought to be very loud, but it is extremely weak. 'Soup of a sausage-stick!'

"I begged her to give me the recipe for the soup, and she explained it to me thus: —

" 'Soup of a sausage-stick is but a cant phrase among men, and is differently interpreted. Every one fancies his own interpretation the best, but in sober reality there is nothing in it whatsoever.'

" 'Nothing!' cried I. That was a poser. 'Truth is not always pleasant, but truth is always the best.' So also said the old owl. I considered the matter, and came to the conclusion that when I brought *the best* I brought more than 'soup of a sausage-stick;' and thereupon I hastened homewards, so that I might arrive in good time to bring what is most valuable — THE TRUTH. The mice are an enlightened community, and their king is the cleverest of them all. He can make me his queen for the sake of Truth."

"Thy truth is a falsehood," said the mouse who had not yet had an opportunity of speaking. "I can make the soup, and I will do it."

V.

HOW THE SOUP WAS MADE.

"I HAVE not travelled at all," said the last mouse. "I remained in our own country. It is not necessary to go to foreign lands — one can learn as well at home. I remained there. I have not acquired any information of unnatural beings. I have not eaten information, or

6*   I

conversed with owls. I confined myself to original thoughts. Will some one now be so good as to fill the kettle with water, and put it on? Let there be plenty of fire under it. Let the water boil — boil briskly; then throw the sausage-stick in. Will his majesty the King of the Mice be so condescending as to put his tail into the boiling pot, and stir it about? The longer he stirs it, the richer the soup will become. It costs nothing, and requires no other ingredients — it only needs to be stirred."

" Cannot another do this?" asked the king.

" No," said the mouse. " The effect can only be produced by the royal tail."

The water was boiled, and the King of the Mice prepared himself for the operation, though it was rather dangerous. He stuck his tail out, as mice are in the habit of doing in the dairy, when they skim the cream off the dish with their tails; but he had no sooner popped his tail into the warm steam than he drew it out and sprang down.

" Of course you are my queen," said he; " but we shall wait for the soup till our golden wedding, and the poor in my kingdom will have something to rejoice over in the future."

So the nuptials were celebrated; but many of the mice, when they went home, said, " It could not well be called soup of a sausage-stick, but rather soup of a mouse's tail."

They allowed that each of the narratives was very well told, but the whole might have been better. " I, for instance, would have related my adventures in such and such words . . . ."

These were the critics, and they are always so wise — afterwards.

And this history went round the world. Opinions were divided about it, but the historian himself remained unmoved. And this is best in great things and in small.

The Neck of a Bottle.

.

YONDER, in the confined, crooked streets, amidst
several poor-looking houses, stood a narrow high
tenement, run up of framework that was much misshapen,
with corners and ends awry. It was inhabited by poor
people, the poorest of whom looked out from the garret,
where, outside the little window, hung in the sunshine an
old, dented bird-cage, which had not even a common
cage-glass, but only the neck of a bottle inverted, with a
cork below, and filled with water. An old maid stood
near the open window; she had just been putting some
chickweed into the cage, wherein a little linnet was hop-
ping from perch to perch, and singing until her warbling
became almost overpowering.

"Yes, you may well sing," said the neck of the bottle;
but it did not say this as we should say it, for the neck of
a bottle cannot speak, but it thought so within itself, just
as we human beings speak inwardly.

"Yes, you may well sing, you who have your limbs
entire. You should have experienced, like me, what it
is to have lost your lower part, to have only a neck and
a mouth, and the latter stopped up with a cork, as I have;

then you would not sing. But it is well that somebody
is contented. I have no cause to sing, and I cannot. I
could once, though, when I was a whole bottle. How I
was praised at the furrier's in the wood, when his daughter
was betrothed! Yes, I remember that day as if it were
yesterday. I have gone through a great deal when I
look back. I have been in fire and in water, down in the
dark earth, and higher up than many; and now I am
suspended outside of a bird-cage in the air and sunshine.
It might be worth while to listen to my story; but I do
not speak it aloud, because I cannot."

So it went on thinking over its own history, which
was curious enough: and the little bird poured forth
its strains, and in the street below people walked and
drove, every one thinking of himself, some scarcely
thinking at all; but the neck of the bottle *was* thinking.

It remembered the blazing smelt-furnace at the manu-
factory where it was blown into life. It remembered
even now that it had been extremely warm; that it had
looked into the roaring oven, its original home, and had
felt strongly inclined to spring back into it; but that by
degrees, as it felt cooler, it found itself comfortable
enough where it was, placed in a row with a whole regi-
ment of brothers and sisters from the same furnace, some
of which, however, were blown into champagne bottles,
others into ale bottles; and that made a difference, since
out in the world an ale bottle may contain the costly
LACRYMÆ CHRISTI, and a champagne bottle may be
filled with blacking; but what they were born to every
one can see by their shape, so that noble remains noble
even with blacking in it.

All the bottles were packed up, and our bottle with them. It then little thought that it would end in being only the neck of a bottle serving as a bird's glass — an honourable state of existence truly, but still something. It did not see daylight again until it was unpacked along with its comrades in the wine merchant's cellar, and was washed for the first time. That was a funny sensation. After that it lay empty and uncorked, and felt so very listless; it wanted something, but did not know what it wanted. At length it was filled with an excellent, superior wine, and, when corked and sealed, a label was stuck on it outside with the words, " Best quality." It was as if it had taken its first academic degree. But the wine was good, and the bottle was good. The young are fond of music, and much singing went on in it, the songs being on themes about which it scarcely knew anything — the green sunlit hills where the wine grapes grew, where beautiful girls and handsome swains met, and danced, and sang, and loved. Ah! there it is delightful to dwell. And all this was made into songs in the bottle, as it is made into songs by young poets, who also frequently know nothing at all about the subjects they choose.

One morning it was bought. The furrier's boy was ordered to purchase a bottle of the best wine, and this one was carried away in a basket, with ham, cheese, and sausage; there were also the nicest butter and the finest bread. The furrier's daughter herself packed the basket. She was so young, so pretty! Her brown eyes laughed, and the smile on her sweet mouth was almost as expressive as her eyes. She had beautiful soft hands

— they were so white; yet her throat and neck were still whiter. It could be seen at once that she was one of the prettiest girls in the neighbourhood, and, strange to say, not yet engaged.

The basket of provisions was placed in her lap when the family drove out to the wood. The neck of the bottle stuck out above the parts of the white napkins that were visible. There was red wax on its cork, and it looked straight into the eyes of the pretty girl, and also into those of the young sailor—the mate of a ship— who sat beside her. He was the son of a portrait painter, and had just passed a first-rate examination for mate, and was to go on board his vessel the next day to sail for far-distant countries. Much was said about his voyage during the drive; and when *it* was spoken of, there was not exactly an expression of joy in the eyes and about the mouth of the furrier's daughter.

The two young people wandered away into the green wood. They were in earnest conversation. Of what were they speaking? The bottle did not hear that, for it was still standing in the basket of provisions. It seemed a long time before it was taken out, but then it saw pleasant faces round. Everybody was smiling, and the furrier's daughter also smiled; but she spoke less, and her cheeks were blushing like two red roses.

The father took the full bottle and the corkscrew. Oh! it is astonishing to a bottle the first time the cork is drawn from it. The neck of the bottle could never afterwards forget that important moment when, with a low sound, the cork flew, and the wine streamed out into the awaiting glasses.

"To the health of the betrothed pair!" cried the father, and every glass was drained; and the young mate kissed his lovely bride. "May happiness and every blessing attend you both!" said the old people; and the young man begged them to fill their glasses again for his toast.

"To my return home and my wedding, within a year and a day!" he cried; and when the glasses were empty he took the bottle, and lifted it high above his head. "Thou hast been present during the happiest day of my life; thou shalt never serve another!"

And he cast the bottle high up in the air. Ah! little did the furrier's daughter think then that she should often look on that which was flung up; but she was destined to do so. It fell among the thick mass of reeds that bordered a pond in the woods. The neck of the bottle remembered distinctly what it thought as it lay there, and it was this: "I gave them wine, and they give me bogwater; but it was well meant." It could no more see the betrothed young couple, or the happy old people; but it heard in the distance the sounds of music and of mirth. Then came two little peasant children peering among the reeds. They saw the bottle, and carried it off with them: so it was provided for.

At home, in the cottage among the woods where they lived, their eldest brother, who was a sailor, had, the day before, come to say farewell; for he was about to start on a long voyage. The mother was busy packing various little matters, which the father was to take with him to the town in the evening, when he went to see his son once more before his departure, and give him again his mother's blessing. A phial with spiced brandy was placed in the package; but at that moment the children came in with the larger, stronger bottle which they had found. A larger quantity could go into it than into the phial. It was not the red wine, as before, that the bottle received, but some bitter stuff. However, it also was excellent as a stomachic. Our bottle was thus again to set forth on its travels. It was carried on board to Peter Jensen, who happened to be in the same ship as was the young mate; but he did not see the bottle, and, if he had seen it, he would not have known it to have been the same from which were drunk the toasts in honour of his betrothal, and to his safe return.

Although there was no longer wine in it, there was something quite as good; and whenever Peter Jensen brought it forth, his comrades called it "the apothecary." The nice medicine was so much in vogue that very soon there was not a drop of it left. The bottle had a pleasant time of it, upon the whole, while its contents were in such high favour. It acquired the name of the great "Lœrke"—"Peter Jensen's Lœrke."*

But this time was passed, and it had lain long neglected in a corner. It did not know whether it was on the voyage out or homewards; for it had never been on shore anywhere. One day a great storm arose; the black, heavy waves rolled mountains high, and heaved the ship up and cast it down by turns; the mast came down with a crash; the sea stove in a plank; the pumps

---

* "Lœrke," which generally means "lark," is the name given among the lower classes in Denmark to a spirit bottle of a peculiar shape. There is no word that corresponds with it in English. — *Trans.*

were no longer of any avail. It was a pitch-dark night.
The ship sank; but at the last minute the young mate
wrote on a slip of paper, "*In the name of Jesus — we
are lost!*" He wrote down the name of his bride, his
own name, and that of his ship; then he thrust the note
into an empty bottle that was within reach, pressed in the
cork tightly, and cast the bottle out into the raging sea.
Little did he know that it was the identical bottle which
had contained the wine in which had been drunk the
toasts of joy and hope for him and her, that was now
tossing on the billows with these last remembrances, and
the message of death.

The ship sank — the crew sank — but the bottle
skimmed the waves like a sea-fowl. It had a heart then
— the letter of love within it. And the sun rose, and
the sun set. This sight recalled to the bottle the scene
of its earliest life — the red glowing furnace, to which it
had once longed to return. It encountered calms and
storms; but it was not dashed to pieces against any rocks.
It was not swallowed by any shark. For more than a
year and a day it drifted on — now towards the north,
now towards the south — as the currents carried it. In
other respects it was its own master; but one can become
tired even of that.

The written paper — the last farewell from the bridegroom to his bride — would only bring deep sorrow if it
ever reached the proper hands. But where were these
hands, that had looked so white when they spread the
table-cloth on the fresh grass in the green wood on the
betrothal-day? Where was the furrier's daughter?
Nay, where was her country? and to what country was

it nearest? The bottle knew not. It drifted and drifted,
and it was so tired of always drifting on; but it could not
help itself. Still, still it had to drift, until at last it
reached the land; but it was a foreign country. It did
not understand a word that was said, for the language
was not such as it had been formerly accustomed to hear;
and one feels quite lost if one does not understand the
language spoken around.

The bottle was taken up and examined; the slip of
paper in it was observed, taken out, and opened; but nobody could make out what was written on it, though
every one knew that the bottle must have been cast overboard, and that some information was contained in the
paper; but what *that* was remained a mystery, and it was
put back into the bottle, and the latter laid by in a large
press, in a large room, in a large house.

Whenever any stranger came the slip of paper was
taken out, opened, and examined, so that the writing,
which was only in pencil, became more and more illegi-
ble from the frequent folding and unfolding of the paper,
till at length the letters could no longer be discerned.
After the bottle had remained about a year in the press
it was removed to the loft, and was soon covered with
dust and cobwebs. Ah! then it thought of its better
days, when red wine was poured from it in the shady
wood, and when it swayed about upon the waves, and
had a secret to carry — a letter, a farewell sigh.

It now remained in the loft for twenty mortal years,
and it might have remained longer, had not the house
been going to be rebuilt. The roof was taken off, the
bottle discovered and talked about; but it did not under-

stand what was said. One does not learn languages,
living up alone in a loft, even in twenty years. " Had I
but been down in the parlour," it thought, and with truth,
" I would, of course, have learned it."

It was now washed and rinsed. It certainly wanted
cleaning sadly, and very clear and transparent it felt
itself after it — indeed, quite young again in its old age ;
but the slip of paper committed to its charge, that was
lost in the washing. The bottle was now filled with
seeds. Such contents were new to it. Well stopped up
and wrapped up it was, and it could see neither a lantern
nor a candle, not to mention the sun or the moon. " One
ought to see something when one goes on a journey,"
thought the bottle ; but it did not, however, until it
reached the place it was going to, and was there unpacked.

" What trouble these people abroad have taken about
it !" was remarked ; " yet no doubt it is cracked." But
it was not cracked. The bottle understood every word
that was said, for they were spoken in the language it
had heard at the furnace, at the wine merchant's, in the
wood, and on board ship — the only right good old language, one which could be understood. The bottle had
returned to its own country, and in its joy had nearly
jumped out of the hands that were holding it. It scarcely
observed that the cork had been removed, its contents
shaken out, and itself put away in the cellar to be kept
and forgotten. But home is dearest, even in a cellar.
It had enough to think over, and time enough to think,
for it lay there for years ; but at last one day folks came
down there to look for some bottles, and took this one
with them.

Outside, in the garden, there were great doings; coloured lamps hung in festoons; paper lanterns, formed like large tulips, gave forth their subdued light. It was also a charming evening; the air was calm and clear; the stars began, one after the other, to shine in the deep blue heavens above ; while the round moon looked like a pale bluish-grey ball, with a golden border encircling it.

There were also some illuminations in the side walks, at least enough to let people see their way ; bottles with lights in them were placed here and there among the hedges ; and amidst these stood the bottle we know, the one that was destined to end as the mere neck of a bottle and the glass of a bird-cage. At the period just named, however, it found everything so exquisitely charming. It was again among flowers and verdure, again surrounded by joy and festivity; it again heard singing and musical instruments, and the hum and buzz of a crowd of people, especially from that part of the gardens which were most brilliantly illuminated. It had a good situation itself, and stood there useful and happy, bearing its appointed light. During such a pleasant time it forgot the twenty years up in the loft, and it is good to be able to forget.

Close by it passed a couple arm-in-arm, like the happy pair in the wood, the mate and the furrier's daughter. It seemed to the bottle as if it were living that time over again. Guests and visitors of different ages wandered up and down, gazing upon the illuminations ; and among these was an old maid, without relations, but not without friends. Probably her thoughts were occupied, as were those of the bottle; for she was thinking of the green

woods, and of a young couple just betrothed. These
souvenirs affected her much, for she had been a party in
them — a prominent party. This was in her happier
hours; and one never forgets these, even when one becomes a very old maid. But she did not recognise the
bottle, and it did not recognise her. So it is we wear
out of each other's knowledge in this world, until people
meet again as these two did.

The bottle passed from the public gardens to the wine
merchant's; it was there again filled with wine, and sold
to an aëronaut, who was to go up in a balloon the following Sunday. There was a multitude of people to witness the ascent, there was a regimental band, and there
were many preparations going on. The bottle saw all
this from a basket, in which it lay with a living rabbit,
who was very much frightened when it saw it was to go
up in the parachute. The bottle did not know where it
was to go; it beheld the balloon extending wider and
wider, and becoming so large that it could not be larger;
then lifting itself up higher and higher, and rolling rest-
lessly until the ropes that held it were cut, when it arose
majestically into the air, with the aëronaut, the basket,
the bottle, and the rabbit; then the music played loudly,
and the assembled crowd shouted, " Hurra! hurra!"

" It is droll to go aloft," thought the bottle; " it is a
novel sort of a voyage. Up yonder one cannot run
away."

Many thousand human beings gazed up at the balloon,
and the old maid gazed among the rest. She stood by
her open garret window, where a cage hung with a little
linnet, which at that time had no water-glass, but had to

content itself with a cup. Just within the window stood
a myrtle tree, that was moved a little aside, that it might
not come in the way while the old maid was leaning out
to look at the balloon. And she could perceive the aëronaut in it; she saw him let the rabbit down in the parachute, and then, having drunk the health of the crowd
below, throw the bottle high up in the air. Little did
she think that it was just the same bottle she had seen
thrown up high in honour of herself and her lover, on a
well-remembered happy day amidst the green wood, when
she was young.

The bottle had no time to think, it was so unexpectedly
exalted to the highest position it had ever attained in its
life. The roofs and the spires lay far below, and the people looked as small as pigmies.

It now descended, and that at a different rate of speed
from the rabbit. The bottle cast somersaults in the air
— it felt itself so young, so buoyant. It was half full of
wine, but not long. What a trip that was! The sun
shone upon the bottle, and all the crowd looked up at it.
The balloon was soon far away, and the bottle was soon
also out of sight, for it fell upon a roof and broke in two;
but the fragments rebounded again, and leaped and rolled
till they reached the yard below, where they lay in smaller
pieces; for only the neck of the bottle escaped destruc-
tion, and it looked as if it had been cut round by a
diamond.

"It may still serve as a glass for a bird's cage," said
the man in the cellar.

But he himself had neither a bird nor a cage, and it
would have cost too much to buy these because he had

found the neck of a bottle that would answer for a glass. The old maid, however, up in the garret, might make use of it; and so the neck of the bottle was sent up to her. A cork was fitted to it, and, as first mentioned, after its many changes, it was filled with fresh water, and was hung in front of the cage of the little bird, that sang until its warbling became almost overpowering.

"Yes, you may well sing," was what the neck of the bottle had said.

It was somewhat of a wonder, as it had been up in a balloon; but with more of its history no one was acquainted. Now it hung as a bird's glass, it could hear the people driving and walking in the street below, and it could hear the old maid talking in her room to a female friend of her youthful days. They were chatting together, but speaking of the myrtle plant in the window, not of the neck of the bottle.

"You must not throw away two rix dollars for a wedding bouquet for your daughter," said the old maid. "You shall have one from me full of flowers. Look how pretty that plant is! Ah! it is a slip of the myrtle tree you gave me the day after my betrothal, that I myself, when the year was past, might take my wedding bouquet from it. But that day never came. The eyes were forever closed that were to have illumined for me the path of happiness in this life. Away, down in the ocean's depths, he sleeps calmly — that angel soul! The tree became an old tree, but I have become still older; and when it died, I took its last green branch and planted it in the earth. That slip has now grown into a high plant, and will at last appear amidst bridal array, and form a wedding bouquet for my friend's daughter."

And tears started to the old maid's eyes. She spoke of the lover of her youth — of the betrothal in the wood; she thought of the toasts that were there drunk; she thought of the first kiss, but she did not speak of that, for she was now but an old maid. She thought of much — much; but little did she think that outside of her window was even then a *souvenir* from that regretted time — the neck of the very bottle that had been drawn when the unforgotten toasts were drunk! Nor did the bottle-neck know her; for it had not heard all she had said, because it had been thinking only of itself.

7   J

The Old Bachelor's Nightcap.

THERE is a street in Copenhagen which bears the extraordinary name of "Hyskenstrœde." And why is it so called? and what is the meaning of that name? It is German; but the German has been corrupted. "Häuschen" it ought to be called, and that signifies "small houses." Those which stood there formerly — and, indeed, for several years — were not much larger than the wooden booths that we see now-a-days erected at fairs. Yes, only a little larger, and with windows; but the panes were of horn or stretched bladder, for in these days it was too expensive to have glass windows in all houses; but the time in question was so far back that our grandfathers' grandfathers, when they mentioned it, also spoke of it as "in ancient days," for it was several hundred years ago.

Many rich merchants in Bremen and Lubeck carried on business in Copenhagen. They did not, however, go there themselves — they sent their clerks; and these persons generally resided in the wooden houses in the "Small Houses' Street," and held sales of ale and spices. The German ale was so excellent, and there were so

many kinds — " Bremer, Prysing, Emser ale," even
" Brunswick Mumme ; " also, all sorts of spices, such as
saffron, anise, ginger, and especially pepper, that was the
most valued; and from this the German commercial travellers acquired the name in Denmark of " Pepper Swains,
or Bachelors." They entered into an agreement before
they left home not to marry; and many of them lived
there to old age. They had to do entirely for themselves, attend to all little domestic matters, even make
their own fires if they had any. Several of them became lonely old men, with peculiar thoughts and peculiar
habits. Every unmarried man who has arrived at a certain age is now here called after them in derision, " Pe-
bersvend" — old bachelor. It was necessary to relate all
this, in order that our story might be understood.

People made great fun of these old bachelors; laughed
at their nightcaps, at their drawing them down over their
eyes, and so retiring to their couches.

> " Saw the firewood, saw it through !
> Old bachelors, there's work for you.
> To bed with you your nightcaps go;
> Put out your lights, and cry, ' O woe ! ' "

Yes, such songs were made on them. People ridiculed
the old bachelor and his nightcap, just because they knew
so little about him, or it. Alas ! let no one desire such a
nightcap. And why not? Listen !

Over in the " Small Houses' Street," in ancient days,
there was no pavement; people stepped from hole to
hole as in a narrow, cut-up defile; and narrow enough
this was, too. The dwellings on the opposite side of the
street stood so close together, that in summer a sail was

spread across the street from one booth to another, and
the whole place was redolent of pepper, saffron, ginger,
and various spices. Behind the desks stood few young
men; no, they were almost all old fellows; and they
were by no means, as we would represent them, crowned
with a peruke or a nightcap, and equipped in shaggy
pantaloons, a vest and coat buttoned tightly up. This
was the costume in which our forefathers were painted,
it is true; but this community of old bachelors could not
afford to have their pictures taken. Yet it would have
been worth while now to have preserved a portrait of one
of them, as they stood behind their desks, or on festival
days, when they wended their way to church. The hat
they wore was broad-brimmed, and with a high crown;
and sometimes one of the younger men would stick a
feather in his. The woollen shirt was concealed by a
deep linen collar; the tight-fitting jacket was closely buttoned, a loose cloak over it; and the pantaloons descended
almost into the square-toed shoes, for stockings they wore
none. In the belt were stuck the eating knife and the
spoon; and, moreover, a large knife as a weapon of defence, for such was often needed in these days.

Thus was equipped, on grand occasions, old Anthon,
one of the oldest bachelors of the " small houses;" only
he did not wear the high-crowned hat, but a fur cap, and
under that a knitted cap, a veritable nightcap, to which
he had so accustomed himself that it was never off his
head: he actually possessed two of the same description.
He would have made an excellent subject for a painter;
he was so skinny, so wrinkled about the mouth and the
eyes; had long fingers, with such large joints; and his

grey eyebrows were so thick. A bunch of grey hair
from one of these hung over his left eye: it certainly was
not pretty, but it made him very remarkable. It was
known that he came from Bremen, at least that his master lived there; but he himself was from Thüringen,
from the town of Eisenach, close to Wartburg. Old
Anthon spoke little of his native place, but he thought of
it the more.

The old lodgers in the street did not associate much
with each other. Each remained in his own booth,
which was locked early in the evening, and then looked
very dismal; for only a glimmering light could be seen
through the horn panes of the window in the roof, beneath which sat, most frequently on his bed, the old man
with his German psalm-book, and chanted the evening
hymn, or else he went out and strolled about at night by
way of amusement; but amusement it could hardly be
called. To be a stranger in a foreign country is a very
sad situation. No notice is taken of him unless he stands
in anyone's way.

Often when it was a pitch-dark night, with pouring
rain, all around looked woefully gloomy and desolate.
No lanterns were to be seen, except the little one that
hung at one end of the street, before the image of the
Virgin Mary that adorned the wall there. The water
was heard dashing and splashing against the wooden
work near, out by Slotsholm, on which the other end of
the street opened. Such evenings are always long and
lonely if there be nothing to interest one. It is not necessary every day to pack and unpack, to make up par-
cels, and to polish scales; but one must have something

to do, and accordingly old Anthon industriously mended
his clothes and cleaned his shoes. When at length he
retired to rest, it was his custom to keep on his nightcap.
At first he would draw it well down, but he would soon
push it up again to look if the light were totally extinguished; nor would he be satisfied without getting up
and feeling it. He would then lie down again, and turn
on the other side, and again draw down the nightcap;
but soon the idea would cross his mind that possibly the
coals might not have become cold in the little fire-pot beneath — the fire might not be totally out — that a spark
might be kindled, fly forth, and do mischief; and he
would get out of his bed and creep down the ladder, for
it could not be called the stairs; and when, on reaching
the fire-pot, he perceived that not a spark was visible,
and he might retire to rest in peace, he would stop half
way up, being seized with the fear that the iron bolt
might not be properly drawn across the door, or the shutters properly secured; and down he would go again,
wearying his poor thin legs. By the time he crept back
to his humble couch he would be half frozen, and his
teeth would be chattering in his head with the cold.
Then he would draw the covering higher up around him,
and his nightcap lower down over his eyes, and his
thoughts would wander from the business and burdens
of the day; but ah! not to soothing scenes. His rev-
eries were never fraught with pleasure, for then came
old reminiscences, and hung their curtains up; and sometimes they were full of pins, that pricked so severely as
to bring tears into his eyes. Such wounds old Anthon
often received, and his warm tears fell on the coverlet or

the floor, sounding as if one of sorrow's deepest strings
had burst; they did not dry up, but kindled into a flame,
which cast its light for him on the panorama of a life —
a picture which never vanished from his mind. Then he
would dry his eyes with his nightcap, and chase away
the tears, and endeavour to chase away the picture with
them; but it would not go, for it was imbedded in his
heart. The panorama did not follow the exact order of
events; also the saddest parts were generally most prominent. And what were these?

" Beautiful are the beech groves in Denmark," it is
said; but still more beautiful did the beech trees in the
meadows near Wartburg seem to Anthon. Mightier and
more majestic seemed to him the old oak up at the proud
baronial castle, where the swinging lantern hung over
the dark masses of rock; sweeter was the perfume of
the apple blossoms there than in the Danish land; he
seemed to feel the charming scent even now. A tear
trickled down his cheeks, and he saw two little children,
a boy and a girl, playing together. The boy had rosy
cheeks, yellow waving hair, and honest blue eyes — he
was the rich merchant's son, little Anthon himself. The
little girl had dark hair and eyes, and she looked bold
and clever — she was the burgomaster's daughter Molly.
The childish couple were playing with an apple. At
length they divided it in two, and each took a half.
They also divided the seeds between them, and ate them
all to one; and the little girl proposed to plant that in
the ground.

" You will see what will come of this — something
will come which you can hardly fancy. An apple tree
will come up, but not all at once."

And they planted the seed in a flower-pot : both of them were very eager about it. The boy dug a hole in the mould with his finger ; the little girl placed the seed in it, and both of them filled up the hole with earth.

" You must not pull it up to-morrow to see if it has taken root," she said ; " that should not be done. I did that with my flower : twice I took it up to see if it was growing. I had very little sense then, and the flower died."

The flower-pot was left in Anthon's care, and every morning, the whole winter through, he looked at it ; but nothing was to be seen except the black earth. Then came spring ; the sun shone so warmly, and two tiny green leaves at last made their appearance in the flower-pot.

" These are Molly and me," said Anthon. " They are charming — they are lovely."

Soon there came a third leaf. Who did that represent ? And leaf after leaf came up ; while day by day, and week by week, the plant became larger and stronger, until it grew into quite a tree. And another tear fell again from its fountain — from old Anthon's heart.

There stretched out, near Eisenach, a range of stony hills, one of which, round in shape, was very conspicuous ; neither tree, nor bush, nor grass grew on it. It was named Mount Venus. Therein dwelt Venus, a goddess from the heathen ages. She was here called Fru Holle, and she knew and could see every child in Eisenach. She had decoyed into her power the noble knight Tannhäuser, the minnesinger, from the musical circle of Wartburg.

Little Molly and Anthon often went to this hill, and she one day said to him, —

"Would you dare to knock on the side of the hill and cry, 'Fru Holle! Fru Holle! open the gate; here is Tannhäuser?' But Anthon dared not do it. Molly dared, however; yet only these words — "Fru Holle! Fru Holle!" — did she say very loudly and distinctly — the rest seemed to die away on the wind; and she certainly did pronounce the rest of the sentence so indistinctly, that Anthon was sure she had not really added the other words. Yet she looked very confident — as bold as when, in the summer evening, she and several other little girls came to play in the garden with him, and when they all wanted to kiss him, just because he would not be kissed, and defended himself from them, she alone ventured to achieve the feat.

"*I* dare to kiss him!" she used to say, with a proud toss of her little head. Then she would take him round his neck to prove her power, and Anthon would put up with it, and think it all right from her. How pretty and how clever she was! Fru Holle within the hill was also very charming, but her charms, it had been said, sprung from the seducing beauty bestowed on her by the evil one; but still greater beauty was to be found in the holy Elizabeth, the patron saint of the country, the pious Thüringian princess, whose good works, known through traditions and legends, were celebrated in so many places. A picture of her hung in the chapel with a silver lamp before it, but Molly did not resemble her.

The apple tree the two children had planted grew year after year; it became so large that it had to be

7 *

transferred to the garden, out in the open air, where the
dew fell and the sun shone warmly; it became strong
enough to withstand the severity of winter, and after
winter's hard trials it seemed as if rejoicing in the return
of spring: it then put forth blossoms.   In August it had
two apples, one for Molly and one for Anthon: it would
not have been well if it had had less.

   The tree had grown rapidly, and Molly had grown as
fast as the tree; she was as fresh as an apple blossom,
but she was no longer to see that flower.   Everything
changes in this world.   Molly's father left his old home,
and Molly went with him — far, far away.   In our time
it might be only a few hours' journey by railway, but in
those days it took more than a day and a night to arrive
so far east from Eisenach.   It was to the other extremity of Thüringia they had to go, to a town which is now
called Weimar.

   And Molly wept, and Anthon wept.   All these were
now concentrated in one single tear, and it had the happy
rosy tinge of joy.   Molly had assured him that she
cared much more for him than for all the grandeur of
Weimar.

   One year passed on, two passed, and a third followed,
and in all that time there came only two letters.   One
was brought by the carrier, the other by a traveller, who
had taken a circuitous course, besides visiting several
cities and other places.

   How often had not Anthon and Molly heard together
the story of Tristand and Isolde, and how often did not
Anthon think of himself and Molly as them !   Although
the name " Tristand " signified that he was born to sor-

row, and that did not apply to Anthon, he never thought,
as Tristand did, " She has forgotten me !" But Isolde
had not forgotten her heart's dear friend ; and when they
were both dead and buried, one on each side of the church,
two linden trees grew out of their graves, and, stretching
over the roof of the church, met there in full bloom.
This was very delightful, thought Anthon, and yet so
sad ! But there could be no sadness where he and
Molly were concerned. And then he whistled an air
of the Minnesinger's "Walther von der Vogelweide," —

> " Under the lime tree by the hedge;''

and especially that favourite verse, —

> " Beyond the wood, in the quiet dale,
> Tandaradni,
> Sang the melodious nightingale.''

This song was always on his lips. He hummed it, and
he whistled it on the clear moonlight night, when, passing
on horseback through the deep ravine, he rode in haste
to Weimar to visit Molly. He wished to arrive unexpectedly, and he *did* arrive unexpectedly.

He was well received. Wine sparkled in the goblets ;
there was gay society, distinguished society. He had a
comfortable room and an excellent bed ; and yet he found
nothing as he had dreamt and thought to find it. He did
not understand himself; he did not understand those
about him ; but we can understand all. One can be in
a house, can mingle with a family, and yet be a total
stranger. One may converse, but it is like conversing
in a stage coach ; may know each other as people know
each other in a stage coach ; be a restraint upon each

other ; wish that one were away, or that one's good neighbour were away ; and it was thus that Anthon felt.

"I will be sincere with you," said Molly to him. "Things have changed much since we were together as children — changed within and without. Habit and will have no power over our hearts. Anthon, I do not wish to have an enemy in you when I am far away from this, as I soon shall be. Believe me, I have a great regard for you ; but to love you — as I now know how one can love another human being — that I have never done. You must put up with this. Farewell, Anthon ! "

And Anthon also said farewell. No tears sprang to his eyes, but he perceived that he was no longer Molly's friend. If we were to kiss a burning bar of iron, or a frozen bar of iron, we should experience the same sensation when the skin came off our lips.

Within twenty-four hours Anthon had reached Eisenach again, but the horse he rode was ruined.

"What of that?" cied he. "I am ruined, and I will ruin all that can remind me of her. Fru Holle! Fru Holle! Thou heathenish woman ! I will tear down and smash the apple tree, and pull it up by the roots. It shall never blossom or bear fruit more."

But the tree was not destroyed; he himself was knocked down, and lay long in a violent fever. What was to raise him from his sick bed? The medicine that did it was the bitterest that could be — one that shook the languid body and the shrinking soul. Anthon's father was no longer the rich merchant. Days of adversity, days of trial, were close at hand. Misfortune rushed in like overwhelming billows — it surged into that once wealthy house. His father became a poor man, and sorrow and calamity paralysed him. Then Anthon found that he had something else to think of than disappointed love, or being angry with Molly. He had now to be both father and mother in his desolate home. He had to arrange everything, look after everything, and to go forth into the world to work for his own and his parents' bread.

He went to Bremen. There he suffered many privations, and passed many melancholy days; and all that he went through sometimes soured his temper, sometimes saddened him, till strength and mind seemed failing. How different were the world and mankind from what he had fancied them in his childhood! What were now to him Minnesingers' poems and songs? They were gall and wormwood. Yes, this was what he often felt; but there were other times when the songs vibrated to his soul, and his mind became calm and peaceful.

"What God wills is always the best," said he then. "It was well that our Lord did not permit Molly's heart to hang on me. What could it have led to, now that prosperity has left me and mine? She gave me up before she knew or dreamed of this reverse from more fortunate days which was hanging over us. It was the mercy of our Lord towards me. Everything is ordained for the best. Yes, all happens wisely. She could not, therefore, have acted otherwise; and yet how bitter have not my feelings been towards her!"

Years passed on. Anthon's father was dead, and stran-

gers dwelt in his paternal home. Anthon, however, was
to see it once more; for his wealthy master sent him on
an errand of business, which obliged him to pass through
his native town, Eisenach. The old WARTBURG stood
unchanged, high up on the hill above, with "the monk
and the nun" in unhewn stone. The mighty oak trees
seemed as imposing as in his childish days. The Venus
mount looked like a grey mass frowning over the valley.
He would willingly have cried, —

"Fru Holle! Fru Holle! open the hill, and let me
stay there, upon the soil of my native home!"

It was a sinful thought, and he crossed himself. Then
a little bird sang among the bushes, and the old Minne-
song came back to his thoughts : —

> " Beyond the wood, in the quiet dale,
> Tandaradai!
> Sang the melodious nightingale."

How remembrances rushed upon him as he approached
the town where his childhood had been spent, which he
now saw through tears! His father's house remained
where it used to be, but the garden was altered; a field
footpath was made across a portion of the old garden;
and the apple tree that he had not uprooted stood there,
but no longer within the garden: it was on the opposite
side of the road, though the sun shone on it as cheerfully
as of old, and the dew fell on it there. It bore such a
quantity of fruit, that the branches were weighed down to
the ground.

"It thrives!" he exclaimed. "Yes, *it* can do so."

One of its well-laden boughs was broken. Wanton

hands had done this, for the tree was now on the side of the public road.

"Its blossoms are carried off without thanks; its fruit is stolen, its branches are broken. It may be said of a tree as of a man, 'It was not sung at the tree's cradle that things should turn out thus.' This one began its life so charmingly; and what has now become of it? Forsaken and forgotten — a garden tree standing in a common field, close to a public road, and bending over a miserable ditch! There it stood now, unsheltered, ill-used, and disfigured! It was not, indeed, withered by all this; but as years advanced, its blossoms would become fewer — its fruit, if it bore any, late; and so it is all over with it."

Thus thought Anthon under the tree, and thus he thought many a night in the little lonely chamber of the wooden house in the "Small Houses' Street," in Copenhagen, whither his rich master had sent him, having stipulated that he was not to marry.

"*He* marry!" He laughed a strange and hollow laugh.

The winter had commenced early. There was a sharp frost, and without there was a heavy snow storm, so that all who could do so kept within doors. Therefore it was that Ánthon's neighbours did not observe that his booth had not been opened for two whole days, and that he had not shown himself during that time. But who would go out in such weather when he could stay at home?

These were dark, dismal days; and in the booth, where the window was not of glass, it looked like twilight, if not sombre night. Old Anthon had scarcely left his bed for two days. He had not strength to get up. The intensely cold weather had brought on a severe fit of rheumatism in his limbs, and the old bachelor lay forsaken and helpless, almost too feeble to stretch out his hand to the pitcher of water which he had placed near his bed; and if he could have done so, it would have been of no avail, for the last drop had been drained from it. It was not the fever, not illness alone, that had thus prostrated him; it was also old age that had crept upon him. It seemed to be constant night up yonder where he lay. A little spider, which he could not see, spun contentedly its gossamer web over his face. It was soon to stretch like a crape veil across the features, when the old man closed his eyes.

He dozed a good deal; yet time seemed long and weary. He shed no tears, and had but little suffering. Molly was scarcely ever in his thoughts. He had a conviction that this world and its bustle were no more for him. At one time he seemed to feel hunger and thirst. He did feel them; but no one came to give him nourishment or drink — no one would come. He thought of those who might be fainting or dying of want. He remembered how the pious Elizabeth, while living on this earth — she who had been the favourite heroine of his childish days at home, the magnanimous Duchess of Thüringia — had herself entered the most miserable abodes, and brought to the sick and wretched refreshments and hope. His thoughts dwelt with pleasure on her good deeds. He remembered how she went to feed the hungry, to speak words of comfort to those who were suffering, and to bind up their wounds, although her austere husband was angry at these works of mercy. He recalled to memory the legend about her, that, as she was going on one of her charitable errands, with a basket well filled with food and wine, her husband, who had watched her steps, rushed out on her, and demanded in high wrath what she was carrying; that, in her fear of him, she replied, " Roses which I have plucked in the garden;" whereupon he dragged the cover off of her basket, and lo! a miracle was worked in favour of the charitable lady, for the wine and bread, and everything in the basket, lay turned into roses.

Thus old Anthon's thoughts wandered to the heroine in history whom he had always so much admired, until her image seemed to stand before his dimming sight, close to his humble pallet in the poor wooden hut in a foreign land.　He uncovered his head, looked in fancy into her mild eyes, and all around him seemed a mingling of lustre and of roses redolent with sweet perfume.　Then he felt the charming scent of the apple blossom, and he beheld an apple tree spreading its blooming branches above him.　Yes, it was the very tree, the seeds of which he and Molly had planted together.　．

And the tree swept its fragrant leaves over his hot brow, and cooled it; they touched his parched lips, and they were like refreshing wine and bread; they fell upon his breast, and he felt himself softly sinking into a calm slumber.

"I shall sleep now," he whispered feebly to himself. "Sleep restores strength—to-morrow I shall be well

and up again. Beautiful, beautiful! The apple tree
planted in love I see again in glory."

And he slept.

The following day — it was the third day the booth
had been shut up — the snow drifted no longer, and the
neighbours went to see about Anthon, who had not yet
shown himself. They found him lying stiff and dead,
with his old nightcap pressed between his hands. They
did not put it upon him in his coffin — he had also
another which was clean and white.

Where now were the tears he had wept? Where
were these pearls? They remained in the nightcap.
Such precious things do not pass away in the washing.
They were preserved and forgotten with the nightcap.
The old thoughts, the old dreams — yes, they remained
still in *the old bachelor's nightcap*. Wish not for that.
It will make your brow too hot, make your pulses beat
too violently, bring dreams that seem reality. This
was proved by the first person who put it on — and that
was not till fifty years after — by the burgomaster himself, who was blessed with a wife and eleven children.
He dreamt of unhappy love, bankruptcy, and short
commons.

"How warm this nightcap is!" he exclaimed, as he
dragged it off. Then pearl after pearl began to fall from
it, and they jingled and glittered. "I must have got the
rheumatism in my head," said the burgomaster. "Sparks
seem falling from my eyes."

They were tears wept half a century before — wept
by old Anthon from Eisenach.

Whoever has since worn that nightcap has sure

enough had visions and dreams; his own history has
been turned into Anthon's; his dream has become quite
a tale, and there were many of them. Let others relate the rest. We have now told the first, and with it
our last words are — Never covet AN OLD BACHELOR'S
NIGHTCAP.

Something.

"I WILL be something," said the oldest of five brothers, "I will be of use in the world, let the position be ever so insignificant which I may fill. If it be only respectable, it will be something. I will make bricks — people can't do without these — and then I shall have done something."

"But something too trifling," said the second brother. "What you propose to do is much the same as doing nothing; it is no better than a hodman's work, and can be done by machinery. You had much better become a mason. *That* is something, and that is what I will be. Yes, that is a good trade. A mason can get into a trade's corporation, become a burgher, have his own colours and his own club. Indeed, if I prosper, I may have workmen under me, and be called 'Master,' and my wife 'Mistress;' and that would be something."

"That is next to nothing," said the third. "There are many classes in a town, and that is about the lowest. It is nothing to be called 'Master.' You might be very superior yourself; but as a master mason you would be only what is called 'a common man.' I know of something better. I will be an architect; enter upon the confines of science; work myself up to a high place in the kingdom of mind. I know I must begin at the foot of the ladder. I can hardly bear to say it— I must begin as a carpenter's apprentice, and wear a cap, though I have been accustomed to go about in a silk hat. I must run to fetch beer and spirits for the common workmen, and let them be 'hail fellow well met' with me. This will be disagreeable; but I will fancy that it is all a masquerade and the freedom of maskers. To-morrow—that is to say, when I am a journeyman—I will go my own way. The others will not join me. I shall go to the academy, and learn to draw and design; then I shall be called an architect. That is something! That is much! I may become 'honourable,' or even 'noble'—perhaps both. I shall build and build, as others have done before me. *There* is something to look forward to—something worth being!"

"But that something I should not care about," said the fourth. "I will not march in the wake of anybody. I will not be a copyist; I will be a genius—will be cleverer than you all put together. I shall create a new style, furnish ideas for a building adapted to the climate and materials of the country—something which shall be a nationality, a development of the resources of our age, and, at the same time, an exhibition of my own genius."

"But if by chance the climate and the materials did not suit each other," said the fifth, "that would be un-fortunate for the result. Nationalities may be so amplified as to become affectation. The discoveries of the age, like youth, may leave you far behind. I perceive right well that none of you will, in reality, become anything, whatever may be your expectations. But do all of you what you please; I shall not follow your examples. I shall keep myself disengaged, and shall reason upon what you perform. There is something wrong in everything. I will pick that out, and reason upon it. That will be something."

And so he did; and people said of the fifth, " He has not settled to anything. He has a good head, but he does nothing."

Even this, however, made him something.

This is but a short history; yet it is one which will not end as long as the world stands.

But is there nothing more about the five brothers? What has been told is absolutely nothing. Hear further; it is quite a romance.

The eldest brother, who made bricks, perceived that from every stone, when it was finished, rolled a small coin; and though these little coins were but of copper, many of them heaped together became a silver dollar; and when one knocks with such at the baker's, the butcher's, and other shops, the doors fly open, and one gets what one wants. The bricks produced all this. The damaged and broken bricks were also made good use of.

Yonder, above the embankment, Mother Margrethe, a poor old woman, wanted to build a small house for herself. She got all the broken bricks, and some whole ones to boot; for the eldest brother had a good heart. The poor woman built her house herself. It was very small; the only window was put in awry, the door was very low, and the thatched roof might have been laid better; but it was at least a shelter and a cover for her. There was a fine view from it of the sea, which broke in its might against the embankment. The salt spray often dashed over the whole tiny house, which still stood there when he was dead and gone who had given the bricks.

The second brother could build in another way. He was also clever in his business. When his apprenticeship was over he strapped on his knapsack, and sang the mechanic's song : —

> " While young, far-distant lands I 'll tread,
>   Away from home to build,
> My handiwork shall win my bread,
>   My heart with hope be filled.
> And when my fatherland I see,
>   And meet my bride — hurra!
> An active workman I shall be:
>   Then who so happy and gay? "

And he *was* that. When he returned to his native town, and became a master, he built house after house — a whole street. It was a very handsome one, and a great ornament to the town. These houses built for him a small house, which was to be his own. But how could the houses build? Ay, ask them that, and they will not answer you; but people will answer for them, and tell you. "It certainly was that street which built him a house." It was only a small one, to be sure, and with a clay floor; but when he and his bride danced on it the floor became polished and bright, and from every stone in the wall sprang a flower which was quite as good as

any costly tapestry.   It was a pleasant house, and they
were a happy couple.   The colours of the masons' com-
pany floated outside, and the journeymen and appren-
tices shouted "Hurra!"   Yes, that was something; and
so he died — and that was also something.

Then came the architect, the third brother, who had
been first a carpenter's apprentice, wearing a cap and
going errands; but, on leaving the academy, rose to be
an architect, and he became a man of consequence.   Yes,
if the houses in the street built by his brother, the master mason, had provided him with a house, a street was
called after the architect, and the handsomest house in it
was his own.   That was something; and he was somebody, with a long, high-sounding title besides.   His children were called people of quality, and when he died his
widow was a widow of rank — that was something.   And
his name stood as a fixture at the corner of the street,
and was often in folks' mouths, being the name of a
street — and that was certainly something.

Next came the genius — the fourth brother — who was
to devote himself to new inventions.   In one of his am-
bitious attempts he fell, and broke his neck; but he had
a splendid funeral, with a procession, and flags, and music.   He was noticed in the newspapers, and three funeral
orations were pronounced over him, the one longer than
the others; and much delighted he would have been with
them if he had heard them, for he was fond of being
talked about.   A monument was erected over his grave.
It was not very grand, but a monument is always something.

He now was dead, as well as the three other brothers;

but the fifth — he who was fond of reasoning or arguing — outlived them all; and that was quite right, for he had thus the last word. And he thought it a matter of great importance to have the last word. It was he who, folks said, "had a good head." At length his last hour also struck. He died, and he arrived at the gate of the kingdom of heaven. Spirits always come there two and two, and along with him stood there another soul, which wanted also to get in, and this was no other than the old Mother Margrethe, from the house on the embankment.

"It must surely be for the sake of contrast that I and yon paltry soul should come here at the same moment," said the reasoner. "Why, who are you, old one? Do you also expect to enter here?" he asked.

And the old woman courtesied as well as she could. She thought it was St. Peter himself who spoke.

"I am a miserable old creature without any family. My name is Margrethe."

"Well, now, what have you done and effected down yonder?"

"I have effected scarcely anything in yonder world — nothing that can tell in my favour here. It will be a pure act of mercy if I am permitted to enter this gate."

"How did you leave yon world?" he asked, merely for something to say. He was tired of standing waiting there.

"Oh! how I left it I really do not know. I had been very poorly, often quite ill, for some years past, and I was not able latterly to leave my bed, and go out into the cold and frost. It was a very severe winter; but I was getting through it. For a couple of days there was a dead

8

calm; but it was bitterly cold, as your honour may remember. The ice had remained so long on the ground,
that the sea was frozen over as far as the eye could reach.
The townspeople flocked in crowds to the ice. I could
hear it all as I lay in my poor room. The same scene
continued till late in the evening — till the moon rose.
From my bed I could see through the window far out beyond the seashore; and there lay on the horizon, just
where the sea and sky seemed to meet, a singular-looking
white cloud. I lay and looked at it; looked at the black
spot in the middle of it, which became larger and larger;
and I knew what that betokened, for I was old and experienced, though I had not often seen that sign. I saw it
and shuddered. Twice before in my life had I seen that
strange appearance in the sky, and I knew that there
would be a terrible storm at the springtide, which would
burst over the poor people out upon the ice, who were
now drinking and rushing about, and amusing themselves.
Young and old — the whole town in fact — were assembled yonder. Who was to warn them of coming danger,
if none of them observed or knew what I now perceived?
I became so alarmed, so anxious, that I got out of my
bed, and crawled to the window. I was incapable of going further; but I put up the window, and, on looking
out, I could see the people skating and sliding and running on the ice. I could see the gay flags, and could
hear the boys shouting hurra, and the girls and the young
men singing in chorus. All was jollity and merriment
there. But higher and higher arose the white cloud with
the black spot in it. I cried out as loud as I could, but
nobody heard me. I was too far away from them. The

wind would soon break loose, the ico give away, and all upon it sink, without any chance of rescue. Hear me they could not, and for me to go to them was impossible. Was there nothing that I could do to bring them back to land? Then our Lord inspired me with the idea of setting fire to my bed; it would be better that my house were to be burned down, than that the many should meet with such a miserable death. Then I kindled the fire. I saw the red flames, and I gained the outside of the house; but I remained lying there. I could do no more, for my strength was exhausted. The blaze pursued me — it burst from the window, and out upon the roof. The crowds on the ice perceived it, and they came running as fast as they could to help me, a poor wretch, whom they thought would be burned in my bed. It was not one or two only who came — they all came. I heard them coming; but I also heard all at once the shrill whistle, the loud roar of the wind. I heard it thunder like the report of a cannon. The springtide lifted the ice, and suddenly it broke asunder; but the crowd had reached the embankment, where the sparks were flying over me. I had been the means of saving them all; but I was not able to survive the cold and fright, and so I have come up here to the gate of the kingdom of heaven; but I am told it is locked against such poor creatures as I. And now I have no longer a home down yonder on the embankment, though that does not insure me any admittance here."

At that moment the gate of heaven was opened, and an angel took the old woman in. She dropped a straw; it was one of the pieces of straw which had stuffed the bed to which she had set fire to save the lives of many, and it had turned to pure gold, but gold that was flexible, and turned itself into pretty shapes.

"See! the poor old woman brought this," said the angel. "What dost thou bring? Ah! I know well; thou hast done nothing — not even so much as making a brick. If thou couldst go back again, and bring only so much as that, if done with good intentions, it would be something: as thou wouldst do it, however, it would be of no avail. But thou canst not go back, and I can do nothing for thee."

Then the poor soul, the old woman from the house on the embankment, begged for him.

"His brother kindly gave me all the stones with which I built my humble dwelling. They were a great gift to a poor creature like me. May not all these stones and fragments be permitted to value as one brick for him? It was a deed of mercy. He is now in want, and this is Mercy's home."

"Thy brother whom thou didst think the most inferior to thyself — him whose honest business thou didst despise — shares with thee his heavenly portion. Thou shalt not be ordered away; thou shalt have leave to remain outside here to think over and to repent thy life down yonder; but within this gate thou shalt not enter until in good works thou hast performed *something*."

"I could have expressed that sentence better," thought the conceited logician; but he did not say this aloud, and that was surely already — SOMETHING.

The Old Oak Tree's Last Dream.

A CHRISTMAS TALE.

THERE stood in a wood, high up on the side of a
sloping hill near the open shore, a very old oak
tree. It was about three hundred and sixty-five years
old, but those long years were not more than as many
single rotations of the earth for us men. We are awake
during the day, and sleep during the night, and have then
our dreams: with the tree it is otherwise. A tree is
awake for three quarters of a year. It only sleeps in
winter — that is *its* night — after the long day which is
called spring, summer, and autumn.

Many a warm summer day had the ephemeron insect
frolicked round the oak tree's head — lived, moved about,
and found itself happy; and when the little creature reposed for a moment in calm enjoyment on one of the
great fresh oak leaves, the tree always said, —

"Poor little thing! one day alone is the span of thy
whole life. Ah, how short! It is very sad."

"Sad!" the ephemeron always replied. "What dost
thou mean by that? Everything is so charming, so warm
and delightful, that I am quite happy."

"But for only one day; then all is over."

"All is over!" exclaimed the insect. "What is the meaning of 'all is over'? Is all over with thee also?"

"No; I may live, perhaps, thousands of thy days, and my lifetime is for centuries. It is so long a period that thou couldst not calculate it."

"No, for I do not understand thee. Thou hast thousands of my days, but I have thousands of moments to be happy in. Is all the beauty in the world at an end when thou diest?"

"Oh! by no means," replied the tree. "It will last longer — much, much longer than I can conceive."

"Well, I think we are much on a par, only that we reckon differently."

And the ephemeron danced and floated about in the sunshine, and enjoyed itself with its pretty little delicate wings, like the most minute flower — enjoyed itself in the warm air, which was so fragrant with the sweet perfumes of the clover-fields, of the wild roses in the hedges, and of the elder-flower, not to speak of the woodbine, the primrose, and the wild mint. The scent was so strong that the ephemeron was almost intoxicated by it. The day was long and pleasant, full of gladness and sweet perceptions; and when the sun set, the little insect felt a sort of pleasing languor creeping over it after all its enjoyments. Its wings would no longer carry it, and very gently it glided down upon the soft blade of grass that was slightly waving in the evening breeze; there it drooped its tiny head, and fell into a calm sleep — the sleep of death.

"Poor little insect!" exclaimed the oak tree, "thy life was far too short."

And every summer's day were repeated a similar dance, a similar conversation, and a similar death. This went on with the whole generation of ephemera, and all were equally happy, equally gay. The oak tree remained awake during its spring morning, its summer day, and its autumn evening; now it was near its sleeping time, its night — the winter was close at hand.

Already the tempests were singing, " Good night, good night! Thy leaves are falling — we pluck them, we pluck them! Try if thou canst slumber; we shall sing thee to sleep, we shall rock thee to sleep; and thy old boughs like this — they are creaking in their joy! Softly, softly sleep! It is thy three hundred and sixty-fifth night. Sleep calmly! The snow is falling from the heavy clouds; it will soon be a wide sheet, a warm coverlet for thy feet. Sleep calmly and dream pleasantly ! "

And the oak tree stood disrobed of all its leaves to go to rest for the whole long winter, and during that time to dream many dreams, often something stirring and exciting, like the dreams of human beings.

It, too, had once been little. Yes, an acorn had been its cradle. According to man's reckoning of time it was now living in its fourth century. It was the strongest and loftiest tree in the wood, with its venerable head reared high above all the other trees; and it was seen far away at sea, and looked upon as a beacon by the navigators of the passing ships. It little thought how many eyes looked out for it. High up amidst its green coronal the wood-pigeons built their nests, and the cuckoo's note was heard from thence; and in the autumn, when the leaves looked like hammered plates of copper, came birds of passage, and rested there before they flew far over the sea. But now it was winter, and the tree stood leafless, and the bended and gnarled branches were naked. Crows and jackdaws came and sat themselves there alternately, and talked of the rigorous weather which was commencing, and how difficult it was to find food in winter.

It was just at the holy Christmas time that the tree dreamt its most charming dream. Let us listen to it.

The tree had a distinct idea that it was a period of some solemn festival; it thought it heard all the church bells round ringing, and it seemed to be a mild summer day. Its lofty head, it fancied, looked fresh and green, while the bright rays of the sun played among its thick foliage. The air was laden with the perfume of wild flowers; various butterflies chased each other in sport around its boughs, and the ephemera danced and amused themselves. All that during years the tree had known and seen around it now passed before it as in a festive procession. It beheld, as in the olden time, knights and ladies on horseback, with feathers in their hats and falcons on their hands, riding through the greenwood; it heard the horns of the huntsmen, and the baying of the hounds; it saw the enemies' troops, with their various uniforms, their polished armour, their lances and halberds, pitch their tents and take them down again; the watch-fires blazed, and the soldiers sang and slept under the sheltering branches of the tree. It beheld lovers meet in the soft moonlight, and cut their names — that first letter — upon its olive-green bark. Guitars and

Æolian harps were again — but there were very many years between them — hung up on the boughs of the tree by gay travelling swains, and again their sweet sounds broke on the stillness around. The wood-pigeons cooed, as if they were describing the feelings of the tree, and the cuckoo told how many summer days it should yet live.

Then it was as if a new current of life rushed from its lowest roots up to its highest branches, even to the farthest leaves; the tree felt that it extended itself therewith, yet it perceived that its roots down in the ground were also full of life and warmth; it felt its strength increasing, and that it was growing taller and taller. The trunk shot up — there was no pause — more and more it grew — its head became fuller, broader — and as the tree grew it became happier, and its desire increased to rise up still higher, even until it could reach the warm, blazing sun.

Already had it mounted above the clouds, which, like multitudes of dark migratory birds, or flocks of white swans, were floating under it; and every leaf of the tree that had eyes could see. The stars became visible during the day, and looked so large and bright: each of them shone like a pair of mild, clear eyes. They might have recalled to memory dear, well-known eyes — the eyes of children — the eyes of lovers when they met beneath the tree.

It was a moment of exquisite delight. Yet in the midst of its pleasure it felt a desire, a longing that all the other trees in the wood beneath — all the bushes, plants, and flowers — might be able to lift themselves

8*   L

like it, and to participate in its joyful and triumphant
feelings. The mighty oak tree, in the midst of its glorious dream, could not be entirely happy unless it had all
its old friends with it, great and small ; and this feeling
pervaded every branch and leaf of the tree as strongly
as if it had lived in the breast of a human being.

The summit of the tree moved about as if it missed
and sought something left behind. Then it perceived
the scent of the woodbine, and soon the still stronger
scent of the violets and wild thyme; and it fancied it
could hear the cuckoo repeat its note.

At length amidst the clouds peeped forth the tops of
the green trees of the wood ; they also grew higher and
higher as the oak had done ; the bushes and the flowers
shot up high in the air; and some of these, dragging
their slender roots after them, flew up more rapidly.
The birch was the swiftest among the trees: like a
white flash of lightning it darted its slender stem upwards, its branches waving like green wreaths and flags.
The wood and all its leafy contents, even the brown-
feathered rushes, grew, and the birds followed them
singing; and in the fluttering blades of silken grass the
grasshopper sat and played with his wings against his
long thin legs, and the wild bees hummed, and all was
song and gladness as up in heaven.

" But the blue-bell and the little wild tansy," said the
oak tree ; " I should like them with me too."

" We are with you," they sang in their low, sweet
tones.

" But the pretty water-lily of last year, and the wild
apple tree that stood down yonder, and looked so fresh,

and all the forest flowers of years past, had they lived
and bloomed till now, they might have been with me."

"We are with you — we are with you," sang their
voices far above, as if they had gone up before.

"Well, this is quite enchanting," cried the old tree.
"I have them all, small and great — not one is forgotten.
How is all this happiness possible and conceivable?"

· "In the celestial paradise all this is possible and conceivable," voices chanted around.

And the tree, which continued to rise, observed that
its roots were loosening from their hold in the earth.

"This is well," said the tree. "Nothing now retains
me. I am free to mount to the highest heaven — to
splendour and light; and all that are dear to me are
with me — small and great — all with me."

"All!"

This was the oak tree's dream; and whilst it dreamt,
a fearful storm had burst over sea and land that holy
Christmas eve. The ocean rolled heavy billows on the
beach — the tree rocked violently, and was torn up by
the roots at the moment it was dreaming that its roots
were loosening. It fell. Its three hundred and sixty-
five years were now as but the day of the ephemeron.

On Christmas morning, when the sun arose, the storm
was passed. All the church bells were ringing joy-
ously; and from every chimney, even the lowest in the
peasant's cot, curled from the altars of the Druidical feast
the blue smoke of the thanksgiving oblation. The sea
became more and more calm, and on a large vessel in
the offing, which had weathered the tempest during the
night, were hoisted all its flags in honour of the day.

"The tree is gone — that old oak tree which was always our landmark!" cried the sailors. "It must have fallen in the storm last night. Who shall replace it? Alas! no one can."

This was the tree's funeral oration — short, but well meant — as it lay stretched at full length amidst the snow upon the shore, and over it floated the melody of the psalm tunes from the ship — hymns of Christmas joy, and thanksgivings for the salvation of the souls of mankind by Jesus Christ, and the blessed promise of everlasting life.

> "Let sacred songs arise on high,
> Loud hallelujahs reach the sky;
> Let joy and peace each mortal share,
> While hymns of praise shall fill the air."

Thus ran the old psalm, and every one out yonder, on the deck of the ship, lifted up his voice in thanksgiving and prayer, just as the old oak tree was lifted up in its last and most delightful dream on that Christmas eve.

The Wind relates the Story of Waldemar Daae and his Daughters.

WHEN the wind sweeps over the grass, it ripples like water; when it sweeps over the corn, it undulates like waves of the sea. All that is the wind's dance. But listen to what the wind tells. It sings it aloud, and it is repeated amidst the trees in the wood, and carried through the loopholes and the chinks in the wall. Look how the wind chases the skies up yonder, as if they were a flock of sheep! Listen how the wind howls below through the half-open gate, as if it were the warder blowing his horn! Strangely does it sound down the chimney and in the fireplace; the fire flickers under it; and the flames, instead of ascending, shoot out towards the room, where it is warm and comfortable to sit and listen to it. Let the wind speak. It knows more tales and adventures than all of us put together. Hearken now to what it is about to relate.

It blew a tremendous blast: that was a prelude to its story.

"There lay close to the Great Belt an old castle with thick red walls," said the wind. "I knew every stone in it. I had seen them before, when they were in Marshal Stig's castle at the Næs. It was demolished. The stones were used again, and became new walls — a new building — at another place, and that was Borreby Castle as it now stands. I have seen and known the high-born ladies and gentlemen, the various generations that have dwelt in it; and now I shall tell about WALDEMAR DAAE AND HIS DAUGHTERS.

"He held his head so high: he was of royal extraction. He could do more than hunt a stag and drain a goblet: that would be proved some day, he said to himself.

"His proud lady, apparelled in gold brocade, walked erect over her polished inlaid floor. The tapestry was magnificent, the furniture costly, and beautifully carved; vessels of gold and silver she had in profusion; there were stores of German ale in the cellars; handsome spirited horses neighed in the stables; all was superb within Borreby Castle when wealth was there.

"And children were there; three fine girls — Idé, Johanné, and Anna Dorthea. I remember their names well even now.

"They were rich people, they were people of distinction — born in grandeur, and brought up in it. Wheugh — wheugh!" whistled the wind; then it continued the tale.

"I never saw there, as in other old mansions, the high-born lady sitting in her boudoir with her maidens and spinning-wheels. She played on the lute, and sang to it, though never the old Danish ballads, but songs in foreign languages.   Here were banqueting and mirth, titled guests came from far and near, music's tones were heard, goblets rang.   I could not drown the noise," said the wind.   "Here were arrogance, ostentation, and display; here was power, but not OUR LORD.

"It was one May-day evening," said the wind.   "I came from the westward.   I had seen ships crushed into wrecks on the west coast of Jutland.   I had hurried over the dreary heaths and green woody coast, had crossed the island of Funen, and swept over the Great Belt, and I was hoarse with blowing.   Then I laid myself down to rest on the coast of Zealand, near Borreby, where there stood the forest and the charming meadows.   The young men from the neighbourhood assembled there, and collected brushwood and branches of trees, the largest and driest they could find.   They carried them to the village, laid them in a heap, and set fire to it; then they and the village girls sang and danced round it.

"I lay still," said the wind; "but I softly stirred one branch — one which had been placed on the bonfire by the handsomest youth.   His piece of wood blazed up, blazed highest.   He was chosen the leader of the rustic game, became 'the wild boar,' and had the first choice among the girls for his 'pet lamb.'   There were more happiness and merriment amongst them than up at the grand house at Borreby.

"And then from the great house at Borreby came, driving in a gilded coach with six horses, the noble lady and her three daughters, so fine, so young — three lovely blossoms — rose, lily, and the pale hyacinth. The mother herself was like a flaunting tulip; she did not deign to notice one of the crowd of villagers, though they stopped their game, and courtesied and bowed with profound respect.

"Rose, lily, and the pale hyacinth — yes, I saw them all three. Whose 'pet lambs' should they one day become? I thought. The 'wild boar' for each of them would assuredly be a proud knight — perhaps a prince. Wheugh — wheugh!

"Well, their equipage drove on with them, and the young peasants went on with their dancing. And. the summer advanced in the village near Borreby, in Tjæreby, and all the surrounding towns.

"But one night when I arose," continued the wind, "the great lady was lying ill, never to move again. That something had come over her which comes over all mankind sooner or later: it is nothing new. Waldemar Daae stood in deep and melancholy thought for a short time. 'The proudest tree may bend, but not break,' said he to himself. The daughters wept; but at last they all dried their eyes at the great house, and the noble lady was carried away; and I also went away," said the wind.

"I returned — I returned soon, over Funen and the Belt, and set myself down by Borreby beach, near the large oak wood. There water-wagtails, wood-pigeons, blue ravens, and even black storks built their nests. It was late in the year: some had eggs, and some had young birds. How they were flying about, and how they were shrieking! The strokes of the axe were heard — stroke after stroke. The trees were to be felled. Waldemar Daae was going to build a costly ship, a man-of-war with three decks, which the king would be glad to pur- chase; and therefore the wood — the seamen's landmark, the birds' home — was to be sacrificed. The great red-backed shrike flew in alarm — his nest was destroyed; the ravens and all the other birds had lost their homes, and flew wildly about with cries of distress and anger. I understood them well. The crows and the jackdaws screamed high in derision. ' From the nest — from the the nest! Away — away!'

"And in the midst of the wood, looking on at the crowd of labourers, stood Waldemar Daae and his three daughters, and they all laughed together at the wild cries of the birds; but his youngest daughter, Anna Dorthea, was sorry for them in her heart; and when the men were about to cut down a partially decayed tree, amidst whose naked branches the black storks had built their nests, and from which the tiny little ones peeped out their heads, she begged it might be spared. She begged — begged with tears in her eyes; and the tree was per- mitted to remain with the nest of black storks. It was not a great boon after all.

"The fine trees were cut down, the wood was sawn, and a large ship with three decks was built. The master shipbuilder himself was of low birth, but of noble ap- pearance. His eyes and his forehead evinced how clever he was, and Waldemar Daae liked to listen to his con- versation; so also did little Idé, his eldest daughter, who was fifteen years of age. And while he was building the ship for the father, he was also building castles in the air for himself, wherein he and Idé sat as man and wife; and that might have happened had the castles been of stone walls, with ramparts and moats, woods and gardens. But, with all his talents, the master shipbuilder was but a humble bird. What should a sparrow do in an eagle's nest?

"Wheugh — wheugh! I flew away, and he flew away, for he dared not remain longer; and little Idé got over his departure, for she was obliged to get over it.

"Splendid dark chargers neighed in the stables, worth being looked at; and they were looked at and admired. An admiral was sent by the king himself to examine the new man-of-war, and to make arrangements for its purchase. He praised the spirited horses loudly. I heard him myself," said the wind. "I followed the gentlemen through the open door, and strewed straw before their feet. Waldemar Daae wanted gold, the admiral wanted the horses — he admired them so much; but the bargain was not concluded, nor was the ship bought — the ship that was lying near the strand, with its white planks — a Noah's ark that was never to be launched upon the deep.

"Wheugh! It was a sad pity.

"In the winter time, when the fields were covered with snow, drift-ice filled the Belt, and I screwed it up to the shore," said the wind. "Then came ravens and crows, all as black as they could be, in large flocks. They perched themselves upon the deserted, dead, lonely ship, that lay high up on the beach; and they cried and lamented, with their hoarse voices, about the wood that was gone, the many precious birds' nests that were laid waste, the old ones rendered homeless, the little ones rendered homeless; and all for the sake of a great lumbering thing, a gigantic vessel, that never was to float upon the deep.

" I whirled the snow in the snow storms, and raised the snow-drifts. The snow lay like a sea high around the vessel. I let it hear my voice, and know what a tempest can say. I knew if I exerted myself it would get some of the knowledge other ships have.

" And winter passed — winter and summer; they come and go as I come and go; the snow melts, the apple blossom blooms, the leaves fall — all is change, change, and with mankind among the rest.

" But the daughters were still young — little Idé a rose, beautiful to look at, as the shipbuilder had seen her. Often did I play with her long brown hair, when, under the apple tree in the garden, she was· standing lost in thought, and did not observe that I was showering down the blossoms upon her head. Then she would start, and gaze at the red sun, and the golden clouds around it, through the space among the dark foliage of the trees.

" Her sister Johanné resembled a lily — fair, slender, and erect; and, like her mother, she was stately and haughty. It was a great pleasure to her to wander up and down the grand saloon where hung the portraits of her ancestors. The high-born dames were painted in silks and velvets, with little hats looped up with pearls on their braided locks — they were beautiful ladies. Their lords were depicted in steel armour, or in costly mantles trimmed with squirrels' fur, and wearing blue ruffs; the sword was buckled round the thigh, and not round the loins. Johanné's own portrait would hang at some future day on that wall, and what would her noble husband be like? Yes, she thought of this, and she said this in low accents to herself. I heard her when I rushed through the long corridor into the saloon, and out again.

"Anna Dorthea, the pale hyacinth, who was only fourteen years of age, was quiet and thoughtful. Her large swimming blue eyes looked somewhat pensive, but a childish smile played around her mouth, and I could not blow it off; nor did I wish to do so.

"I met her in the garden, in the ravine, in the fields. She was gathering plants and flowers, those which she knew her father made use of for the drinks and drops he was fond of distilling. Waldemar Daae was arrogant and conceited, but also he had a great deal of knowledge. Everybody knew that, and everybody talked in whispers about it. Even in summer a fire burned in his private cabinet; its doors were always locked. He passed days and nights there, but he spoke little about his pursuits. The mysteries of nature are studied in silence. He expected soon to discover its greatest secret — the transmutation of other substances into gold.

"It was for this that smoke was ever issuing from the chimney of his laboratory; for this that sparks and flames were always there. And I was there too," said the wind. "'Hollo, hollo!' I sang through the chimney. There were steam, smoke, embers, ashes. 'You will burn yourself up — take care, take care!' But Waldemar Daae did *not* take care.

"The splendid horses in the stables, what became of them? — the silver and the gold plate, the cows in the fields, the furniture, the house itself? Yes, they could be smelted — smelted in the crucibles; and yet no gold was obtained.

"All was empty in the barns and in the pantry, in the cellars and in the loft. The fewer people, the more mice. One pane of glass was cracked, another was broken. I did not require to go in by the door," said the wind. "When the kitchen chimney is smoking, dinner is preparing; but there the smoke rolled from the chimney for that which devoured all repasts — for the yellow gold.

"I blew through the castle gate like a warder blowing his horn; but there was no warder," said the wind. "I turned the weathercock above the tower — it sounded like a watchman snoring inside the tower; but no watchman was there — it was only kept by rats and mice. Poverty presided at the table — poverty sat in the clothes' chests and in the store-rooms. The doors fell off their hinges — there came cracks and crevices everywhere. I went in, and I went out," said the wind; "therefore I knew what was going on.

"Amidst smoke and ashes — amidst anxiety and sleepless nights — Waldemar Daae's hair had turned grey; so had his beard and the thin locks on his forehead; his skin had become wrinkled and yellow, his eyes ever straining after gold — the expected gold.

"I whisked smoke and ashes into his face and beard: debts came instead of gold. I sang through the broken windows and cracked walls — came moaning into the daughter's cheerless room, where the old bed-gear was faded and threadbare, but had still to hold out. Such a song was not sung at the children's cradles. High life

had become wretched life. I was the only one then who
sang loudly in the castle," said the wind. "I showed
them in, and they said they were comfortable. They
had no wood to burn — the trees had been felled from
which they would have got it. It was a sharp frost. I
rushed through loopholes and corridors, over roofs and
walls, to keep up my activity. In their poor chamber
lay the three aristocratic daughters in their bed to keep
themselves warm. To be as poor as church mice — that
was high life! Wheugh! Would they give it up?
But Herr Daae could not.

"'After winter comes spring,' said he. 'After want
come good times; but they make one wait. The castle
is now mortgaged — we have arrived at the worst — we
shall have gold now at Easter!'

"I heard him murmuring near a spider's web : —

"'Thou active little weaver! thou teachest me to per-
severe. Even if thy web be swept away, thou dost commence again, and dost complete it. Again let it be torn
asunder, and, unwearied, thou dost again recommence thy
work over and over again. I shall follow thy example.
I will go on, and I shall be rewarded.'

"It was Easter morning — the church bells were ringing. The sun was careering in the heavens. Under a
burning fever the alchemist had watched all night: he
had boiled and cooled — mixed and distilled. I heard
him sigh like a despairing creature; I heard him pray ;
I perceived that he held his breath in his anxiety. The
lamp had gone out — he did not seem to notice it. I
blew on the red-hot cinders; they brightened up, and
shone on his chalky-white face, and tinged it with a mo-

mentary brightness. The eyes had almost closed in their
deep sockets; now they opened wider — wider — as if
they were about to spring forth.

" Look at the alchemical glass! There is something
sparkling in it! It is glowing, pure, heavy! He lifted
it with a trembling hand. He cried with trembling lips,
' Gold — gold!' He staggered, and seemed quite giddy
at the sight. I could have blown him away," said the
wind; " but I only blew in the ruddy fire, and followed
him through the door in to where his daughters were
freezing. His dress was covered with ashes; they were
to be seen in his beard, and in his matted hair. He
raised his head proudly, stretched forth his rich treasure
in the fragile glass, and ' Won — won! gold!' he cried,
as he held high in the air the glass that glittered in the
dazzling sunshine. But his hand shook, and the alchemical glass fell to the ground, and broke into a thousand
pieces. The last bubble of his prosperity had burst.
Wheugh — wheugh! And I darted away from the alchemist's castle.

" Later in the year, during the short days, when fogs
come with their damp drapery, and wring out wet drops
on the red berries and the leafless trees, I came in a
hearty humour, sent breezes aloft to clear the air, and
began to sweep down the rotten branches. That was no
hard work, but it was a useful one. There was sweeping of another sort within Borreby Castle, where Waldemar Daae dwelt. His enemy, Ové Ramel, from
Basnæs, was there, with the mortgage bonds upon the
property and the dwelling-house, which he had pur-
chased. I thundered against the cracked window-panes,

slammed the rickety doors, whistled through the cracks
and crevices, ' Wheu-gh !' Herr Ové should have no
pleasure in the prospect of living there.   Idé and Anna
Dorthea wept bitterly.   Johanné stood erect and composed; but she looked very pale, and bit her lips till they
bled.   Much good would that do !   Ové Ramel vouchsafed his permission to Herr Daae to remain at the castle during the rest of his days ; but he got no thanks for
the offer.   I overheard all that passed.   I saw the homeless man draw himself up haughtily, and toss his head ;
and I sent a blast against the castle and the old linden
trees, so that the thickest branch among them broke,
though it was not rotten.   It lay before the gate like a
broom, in case something had to be swept out; and to be
sure there *was* a clean sweep.

" It was a sad day, a cruel hour, a heavy trial to sus-
tain; but the heart was hard — the neck was stiff.

" They possessed nothing but the clothes they had on.
Yes, they had a newly-bought alchemist's glass, which
was filled with what had been wasted on the floor: it
had been scraped up, the treasure promised, but not
yielded.   Waldemar Daae concealed this near his breast,
took his stick in his hand, and the once wealthy man
went, with his three daughters, away from Borreby Castle.   I blew coldly on his wan cheeks, and ruffled his
grey beard and his long white hair.   I sang round them,
' Wheu-gh — wheu-gh !'

" There was an end to all their grandeur!

"Idé and Anna Dorthea walked on each side of their
father ; Johanné turned round at the gate.   Why did she

do so? Fortune would not turn. She gazed at the red stones of the wall, the stones from Marshal Stig's castle, and she thought of his daughters: —

> ' The eldest took the younger's hand,
> And out in the wide world they went,'

She thought upon that song. Here there were three, and their father was with them. They passed as beggars over the same road where they had so often driven in their splendid carriage to SMIDSTRUP MARK, to a house with mud floors that was let for ten marks a year — their new manor-house, with bare walls and empty closets. The crows and the jackdaws flew after them, and cried, as if in derision, ' From the nest — from the nest! away — away!' as the birds had screeched at Borreby Wood when the trees were cut down.

"And thus they entered the humble house at Smidstrup Mark, and I wandered away over moors and meadows, through naked hedges and leafless woods, to the open sea — to other lands. Wheugh — wheugh! On — on — on ! "

What became of Waldemar Daae? What became of his daughters? The wind will tell.

" The last of them I saw was Anna Dorthea, the pale hyacinth. She had become old and decrepit: that was about fifty years after she had left the castle. She lived the longest — she saw them all out."

" Yonder, on the heath, near the town of Viborg, stood the dean's handsome house, built of red granite. The smoke rolled plentifully from its chimneys. The gentle

9   M

lady and her beautiful daughters sat on the balcony, and
looked over their pretty garden on the brown heath.   At
what were they gazing?   They were looking at the
storks' nests, on a castle that was almost in ruins.   The
roof, where there was any roof, was covered with moss
and houseleeks; but the best part of it sustained the
storks' nests — that was the only portion which was in
tolerable repair.

"It was a place to look at, not to dwell in.   I had to
be cautious with it," said the wind.   "For the sake of
the storks the house was allowed to stand, else it was
really a disgrace to the heath.   The dean would not
have the storks driven away; so the dilapidated building
was permitted to remain, and a poor woman was permitted to live in it.   She had to thank the Egyptian
birds for that — or was it a reward for having formerly
begged that the nests of their wild black kindred might
be spared in Borreby Wood?   *Then* the wretched pauper
was a young girl — a lovely pale hyacinth in the noble
flower parterre.   She remembered it well — poor Anna
Dorthea!

"'Oh! oh!   Yes, mankind can sigh as the wind does
amidst the sedges and the rushes — Oh!   No church bell
tolled at *thy* death, Waldemar Daae!   No charity-school
children sang over his grave when the former lord of
Borreby was laid in the cold earth!   Oh, all shall come
to an end, even misery!   Sister Idé became a peasant's
wife.   That was the hardest trial to her poor father.
His daughter's husband a lowly serf, who could be
obliged by his master to perform the meanest tasks!   He,
too, is now under the sod, and thou art there with him,

unhappy Idé! O yes — O yes! it was not all over, even then; for I am left a poor, old, helpless creature. Blessed Christ! take me hence!'

"Such was Anna Dorthea's prayer in the ruined castle, where she was permitted to live — thanks to the storks.

"The boldest of the sisters I disposed of," said the wind. "She dressed herself in men's clothes, went on board a ship as a poor boy, and hired herself as a sailor. She spoke very little, and looked very cross, but was willing to work. She was a bad hand at climbing, however; so I blew her overboard before any one had found out that she was a female; and I think that was very well done on my part," said the wind.

"It was one Easter morning, the anniversary of the very day on which Waldemar Daae had fancied that he had found out the secret of making gold, that I heard under the storks' nests, from amidst the crumbling walls, a psalm tune — it was Anna Dorthea's last song.

"There was no window. There was only a hole in the wall. The sun came like a mass of gold, and placed itself there. It shone in brightly. Her eyes closed — her heart broke! They would have done so all the same, had the sun not that morning blazed in upon her.

"The storks had provided a roof over her head until her death.

"I sang over her grave," said the wind; "I had also sung over her father's grave, for I knew where it was, and none else did.

"New times came — new generations. The old highway had disappeared in inclosed fields. Even the tombs, that were fenced around, have been converted into a new road ; and the railway's steaming engine, with its lines of carriages, dashes over the graves, which are as much forgotten as the names of those who moulder into dust in them ! Wheugh — wheugh !

"This is the history of Waldemar Daae and his daughters. Let any one relate it better who can," said the wind, turning round.

And he was gone !

The Girl who Trod upon Bread.

YOU have doubtless heard of the girl who trod upon bread, not to soil her pretty shoes, and what evil this brought upon her. The tale is both written and printed.

She was a poor child, but proud and vain. She had a bad disposition, people said. When she was a little more than an infant, it was a pleasure to her to catch flies, to pull off their wings, and maim them entirely. She used, when somewhat older, to take lady-birds and beetles, stick them all upon a pin, then put a large leaf or a piece of paper close to their feet, so that the poor things held fast to it, and turned and twisted in their endeavours to get off the pin.

"Now the lady-birds shall read," said little Inger. "See how they turn the paper!"

As she grew older she became worse instead of better; but she was very beautiful, and that was her misfortune. She would have been punished otherwise, and in the long run she was.

"You will bring evil on your own head," said her mother. "As a little child you used often to tear my aprons; I fear that when you are older you will break my heart."

And she did so sure enough.

At length she went into the country to wait on people of distinction. They were as kind to her as if she had been one of their own family; and she was so well dressed that she looked very pretty, and became extremely arrogant.

When she had been a year in service her employers said to her, —

"You should go and visit your relations, little Inger."

She went, resolved to let them see how fine she had become. When, however, she reached the village, and saw the lads and lasses gossiping together near the pond, and her mother sitting close by on a stone, resting her head against a bundle of firewood which she had picked up in the forest, Inger turned back. She felt ashamed that she who was dressed so smartly should have for her mother such a ragged creature, one who gathered sticks for her fire. It gave her no concern that she was expected — she was so vexed.

A half year more had passed.

"You must go home some day and see your old parents, little Inger," said the mistress of the house. "Here is a large loaf of white bread — you can carry this to them; they will be rejoiced to see you."

And Inger put on her best clothes and her nice new shoes, and she lifted her dress high, and walked so carefully, that she might not soil her garments or her feet. There was no harm at all in that. But when she came to where the path went over some damp marshy ground, and there were water and mud in the way, she threw the bread into the mud, in order to step upon it and get over with dry shoes; but just as she had placed one foot on the bread, and had lifted the other up, the bread sank in with her deeper and deeper, till she went entirely down, and nothing was to be seen but a black bubbling pool.

That is the story.

What became of the girl? She went below to the *Old. Woman of the Bogs*, who brews down there. The Old Woman of the Bogs is an aunt of the fairies. *They* are very well known. Many poems have been written about them, and they have been printed; but nobody knows anything more of the Old Woman of the Bogs than that, when the meadows and the ground begin to reek in summer, it is the old woman below who is brewing. Into her brewery it was that Inger sank, and no one could hold out very long there. A cesspool is a charming apartment compared with the old Bog-woman's brewery. Every vessel is redolent of horrible smells, which would make any human being faint, and they are packed closely together and over each other; but even if there were a small space among them which one might creep through, it would be impossible, on account of all the slimy toads and snakes that are always crawling and forcing themselves through. Into this place little Inger sank. All this nauseous mess was so ice-cold that she shivered in every limb. Yes, she became stiffer and stiffer. The bread stuck fast to her, and it drew her as an amber bead draws a slender thread.

The Old Woman of the Bogs was at home. The brewery was that day visited by the devil and his dam, and she was a venomous old creature who was never idle. She never went out without having some needle-work with her. She had brought some there. She was sewing running leather to put into the shoes of human beings, so that they should never be at rest. She embroidered lies, and worked up into mischief and discord thoughtless words, that would otherwise have fallen to the ground. Yes, she knew how to sew and embroider, and transfer with a vengeance, that old gran-dam!

She beheld Inger, put on her spectacles, and looked at her.

"That is a girl with talents," said she. "I shall ask for her as a *souvenir* of my visit here; she may do very well as a statue to ornament my great-grand-children's antechamber;" and she took her.

It was thus little Inger went to the infernal regions. People do not generally go straight through the air to them: they can go by a roundabout path when they know the way.

It was an antechamber in an infinity. One became giddy there at looking forwards, and giddy at looking backwards, and there stood a crowd of anxious, pining beings, who were waiting and hoping for the time when the gates of grace should be opened. They would have long to wait. Hideous, large, waddling spiders wove thousands of webs over their feet; and these webs were like gins or foot-screws, and held them as fast as chains of iron, and were a cause of disquiet to every soul —

a painful annoyance. Misers stood there, and lamented that they had forgotten the keys of their money chests. It would be too tiresome to repeat all the complaints and troubles that were poured forth there. Inger thought it shocking to stand there like a statue: she was, as it were, fastened to the ground by the bread.

"This comes of wishing to have clean shoes," said she to herself. "See how they all stare at me!"

Yes, they did all stare at her; their evil passions glared from their eyes, and spoke, without sound, from the corner of their mouths: they were frightful.

"It must be a pleasure to them to see me," thought little Inger. "I have a pretty face, and am well dressed;" and she dried her eyes. She had not lost her conceit. She had not then perceived how her fine clothes had been soiled in the brewhouse of the Old Woman of the Bogs. Her dress was covered with dabs of nasty matter; a snake had wound itself among her hair, and it dangled over her neck; and from every fold in her garment peeped out a toad, that puffed like an asthmatic lap-dog. It was very disagreeable. "But all the rest down here look horrid too," was the reflection with which she consoled herself.

But the worst of all was the dreadful hunger she felt. Could she not stoop down and break off a piece of the bread on which she was standing? No; her back was stiffened; her hands and her arms were stiffened; her whole body was like a statue of stone; she could only move her eyes, and these she could turn entirely round, and that was an ugly sight. And flies came and crept over her eyes backwards and forwards. She winked her

9 *

eyes; but the intruders did not fly away, for they could not — their wings had been pulled off. That was another misery added to the hunger — the gnawing hunger that was so terrible to bear!

"If this goes on I cannot hold out much longer," she said.

But she had to hold out, and her sufferings became greater.

Then a warm tear fell upon her head. It trickled over her face and her neck, all the way down to the bread. Another tear fell, then many followed. Who was weeping over little Inger? Had she not a mother up yonder on the earth? The tears of anguish which a mother sheds over her erring child always reach it; but they do not comfort the child — they burn, they increase the suffering. And oh! this intolerable hunger; yet not to be able to snatch one mouthful of the bread she was treading under foot! She became as thin, as slender as a reed. Another trial was that she heard distinctly all that was said of her above on the earth, and it was nothing but blame and evil. Though her mother wept, and was in much affliction, she still said, —

"Pride goes before a fall. That was your great fault, Inger. Oh, how miserable you have made your mother!"

Her mother and all who were acquainted with her were well aware of the sin she had committed in treading upon bread. They knew that she had sunk into the bog, and was lost; the cowherd had told that, for he had seen it himself from the brow of the hill.

"What affliction you have brought on your mother, Inger!" exclaimed her mother. "Ah, well! I expected no better from you."

"Would that I had never been born!" thought Inger; "that would have been much better for me. My mother's whimpering can do no good now."

She heard how the family, the people of distinction who had been so kind to her, spoke. "She was a wicked child," they said; "she valued not the gifts of our Lord, but trod them under her feet. It will be difficult for her to get the gates of grace open to admit her."

"They ought to have brought me up better," thought Inger. "They should have taken the whims out of me, if I had any."

She heard that there was a common ballad made about her, "the bad girl who trod upon bread, to keep her shoes nicely clean," and this ballad was sung from one end of the country to the other.

"That any one should have to suffer so much for such as that — be punished so severely for such a trifle!" thought Inger. "All these others are punished justly, for no doubt there was a great deal to punish; but ah, how I suffer!"

And her heart became still harder than the substance into which she had been turned.

"No one can be better in such society. I will not grow better here. See how they glare at me!"

And her heart became still harder, and she felt a hatred towards all mankind.

"They have a nice story to tell up there now. Oh, how I suffer!"

She listened, and heard them telling her history as a

warning to children, and the little ones called her "ungodly Inger." "She was so naughty," they said, "so very wicked, that she deserved to suffer."

The children always spoke harshly of her. One day, however, that hunger and misery were gnawing her most dreadfully, and she heard her name mentioned, and her story told to an innocent child — a little girl — she observed that the child burst into tears in her distress for the proud, finely-dressed Inger.

"But will she never come up again?" asked the child.

The answer was, —

"She will never come up again."

"But if she will beg pardon, and promise never to be naughty again?"

"But she will *not* beg pardon," they said.

"Oh, how I wish she would do it!" sobbed the little girl in great distress. "I will give my doll, and my doll's house too, if she may come up! It is so shocking for poor little Inger to be down there!"

These words touched Inger's heart; they seemed almost to make her good. It was the first time any one had said "poor Inger," and had not dwelt upon her faults. An innocent child cried and prayed for her. She was so much affected by this that she felt inclined to weep herself; but she could not, and this was an additional pain.

Years passed on in the earth above; but down where she was there was no change, except that she heard more and more rarely sounds from above, and that she herself was more seldom mentioned. At last one day she heard a sigh, and "Inger, Inger, how miserable you have made me! I foretold that you would!" These were her mother's last words on her deathbed.

And again she heard herself named by her former employers, and her mistress said, —

" Perhaps I may meet you once more, Inger. None know whither they are to go."

But Inger knew full well that her excellent mistress would never come to the place where *she* was.

Time passed on, and on, slowly and wretchedly. Then once more Inger heard her name mentioned, and she beheld as it were, directly above her, two clear stars shining. These were two mild eyes that were closing upon earth. So many years had elapsed since a little girl had cried in childish sorrow over " poor Inger," that that child had become an old woman, whom our Lord was now about to call to himself. At that hour, when the thoughts and the actions of a whole life stand in review before the parting soul, she remembered how, as a little child, she had wept bitterly on hearing the history of Inger. That time, and those feelings, stood so prominently before the old woman's mind in the hour of death, that she cried with intense emotion, —

" Lord, my God! have not I often, like Inger, trod under foot Thy blessed gifts, and placed no value on them? Have I not often been guilty of pride and vanity in my secret heart? But Thou, in Thy mercy, didst not let me sink; Thou didst hold me up. Oh, forsake me not in my last hour!"

And the aged woman's eyes closed, and her spirit's eyes opened to what had been formerly invisible; and as Inger had been present in her latest thoughts, she beheld

her, and perceived how deep she had been dragged down-
wards. At that sight the gentle being burst into tears;
and in the kingdom of heaven she stood like a child, and
wept for the fate of the unfortunate Inger. Her tears
and her prayers sounded like an echo down in the hollow
form that confined the imprisoned, miserable soul. That
soul was overwhelmed by the unexpected love from those
realms afar. One of God's angels wept for her! Why
was this vouchsafed to her? The tortured spirit gathered, as it were, into one thought, all the actions of its
life — all that it had done; and it shook with the violence
of its remorse — remorse such as Inger had never felt.
Grief became her predominating feeling. She thought
that for her the gates of mercy would never open, and as
in deep contrition and self-abasement she thought thus,
a ray of brightness penetrated into the dismal abyss — a
ray more vivid and glorious than the sunbeams which
thaw the snow figures that the children make in their
gardens. And this ray, more quickly than the snow-
flake that falls upon a child's warm mouth can be melted
into a drop of water, caused Inger's petrified figure to
evaporate, and a little bird arose, following the zigzag
course of the ray, up towards the world that mankind
inhabit. But it seemed afraid and shy of everything
around it; it felt ashamed of itself; and apparently
wishing to avoid all living creatures, it sought, in haste,
concealment in a dark recess in a crumbling wall. Here
it sat, and it crept into the farthest corner, trembling all
over. It could not sing, for it had no voice. For a long
time it sat quietly there before it ventured to look out and
behold all the beauty around. Yes, it was beauty! The

air was so fresh, yet so soft; the moon shone so clearly; the trees and the flowers scented so sweetly; and it was so comfortable where she sat — her feather garb so clean and nice! How all creation told of love and glory! The grateful thoughts that awoke in the bird's breast she would willingly have poured forth in song, but the power was denied to her. Yes, gladly would she have sung as do the cuckoo and nightingale in spring. Our gracious Lord, who hears the mute worm's hymn of praise, understood the thanksgiving that lifted itself up in the tones of thought, as the psalm floated in David's mind before it resolved itself into words and melody.

As weeks passed on, these unexpressed feelings of gratitude increased. They would surely find a voice some day, with the first stroke of the wing, to perform some good act. Might not this happen?

Now came the holy Christmas festival. The peasants raised a pole close by the old wall, and bound an unthrashed bundle of oats on it, that the birds of the air might also enjoy the Christmas, and have plenty to eat at that time which was held in commemoration of the redemption brought to mankind.

And the sun rose brightly that Christmas morning, and shone upon the oat-sheaf, and upon all the chirping birds that flew around the pole; and from the wall issued a faint twittering. The swelling thoughts had at last found vent, and the low sound was a hymn of joy, as the bird flew forth from its hiding-place.

The winter was an unusually severe one. The waters were frozen thickly over; the birds and the wild animals in the woods had great difficulty in obtaining food. The little bird, that had so recently left its dark solitude, flew about the country roads, and when it found by chance a little corn dropped in the ruts, it would eat only a single grain itself, while it called all the starving sparrows to partake of it. It would also fly to the villages and towns, and look well about; and where kind hands had strewed crumbs of bread outside the windows for the birds, it would eat only one morsel itself, and give all the rest to the others.

At the end of the winter the bird had found and given away so many crumbs of bread, that the number put together whould have weighed as much as the loaf upon which little Inger had trodden in order to save her fine shoes from being soiled; and when she had found and given away the very last crumb, the grey wings of the bird became white, and expanded wonderfully.

" It is flying over the sea ! " exclaimed the children who saw the white bird. Now it seemed to dip into the ocean, now it arose into the clear sunshine; it glittered in the air; it disappeared high, high above; and the children said that it had flown up to the sun.

Olé, the Watchman of the Tower.

"IN the world it is always going up and down, and down and up again; but I can't go higher than I am," said Olé, the watchman of the church tower. "Ups and downs most people have to experience; in point of fact, we each become at last a kind of tower-watchman — we look at life and things from above."

Thus spoke Olé up in the lofty tower — my friend the watchman, a cheerful, chatty old fellow, who seemed to blurt everything out at random, though there were, in reality, deep and earnest feelings concealed in his heart. He had come of a good stock; some people even said that he was the son of a *Conferentsraad,** or might have been that. He had studied, had been a teacher's assistant, assistant clerk in the church; but these situations had not done much for him. At one time he lived at the chief clerk's, and was to have bed and board free. He was then young, and somewhat particular about his dress, as I have heard. He insisted on having his boots polished and brushed with blacking, but the head clerk

---

* A Danish title.

would only allow grease; and this was a cause of dis-
sension between them. The one talked of stinginess, the
other talked of foolish vanity. The blacking became the
dark foundation of enmity, and so they parted ; but what
he had demanded from the clerk he also demanded from
the world — real blacking ; and he always got its substi-
tute, grease; so he turned his back upon all mankind,
and became a hermit. But a hermitage coupled with a
livelihood is not to be had in the midst of a large city
except up in the steeple of a church. Thither he betook
himself, and smoked his pipe in solitude. He looked up,
and he looked down; reflected according to his fashion
upon all he saw, and all he did not see — on what he
read in books, and what he read in himself.

I often lent him books, good books ; and people can
converse about these, as everybody knows. He did not
care for fashionable English novels, he said, nor for
French ones either — they were all too frivolous. No,
he liked biographies, and books that relate to the wonders
of nature. I visited him at least once a year, generally
immediately after the New Year. He had then always
something to say that the peculiar period suggested to
his thoughts.

I shall relate what passed during two of my visits,
and give his own words as nearly as I can.

THE FIRST VISIT.

Among the books I had last lent Olé was one about
pebbles, and it pleased him extremely.

" Yes, sure enough they are veterans from old days,

these pebbles," said he; "and yet we pass them carelessly by. I have myself often done so in the fields and on the beach, where they lie in crowds. We tread them under foot in some of our pathways, these fragments from the remains of antiquity. I have myself done that; but now I hold all these pebble-formed pavements in high respect. Thanks for that book; it has driven old ideas and habits of thinking aside, and has replaced them by other ideas, and made me eager to read something more of the same kind. The romance of the earth is the most astonishing of all romances. What a pity that one cannot read the first portion of it — that it is composed in a language we have not learned! One must read it in the layers of the ground, in the strata of the rocks, in all the periods of the earth. It was not until the sixth part that the living and acting persons, Mr. Adam and Mrs. Eve, were introduced, though some will have it they came immediately. That, however, is all one to me. It is a most eventful tale, and we are all in it. We go on digging and groping, but always find ourselves where we were; yet the globe is ever whirling round, and without the waters of the world overwhelming us. The crust we tread on holds together — we do not fall through it; and this is a history of a million of years, with constant advancement. Thanks for the book about the pebbles. They could tell many a strange tale if they were able.

"Is it not pleasant once and away to become like a Nix when one is perched so high as I am, and then to remember that we all are but minute ants upon the earth's ant-hill, although some of us are distinguished ants, some are laborious, and some are indolent ants? One seems

to be so excessively young by the side of these million
years old, reverend pebbles. I was reading the book on
New Year's eve, and was so wrapped up in it that I forgot my accustomed amusement on that night, looking at
' the wild host to Amager,' of which you may have
heard.

"The witches' journey on broomsticks is well known
— that takes place on St. John's night, and to Bloksberg.
But we have also the wild host, here at home and in our
own time, which goes to Amager every New Year's eve.
All the bad poets and poetesses, newspaper writers, mu-
sicians, and artists of all sorts, who come before the public, but make no sensation — those, in short, who are very
mediocre, ride — on New Year's eve, out to Amager :
they sit astride on their pencils or quill pens. Steel
pens don't answer, they are too stiff. I see this troop, as
I have said, every New Year's eve. I could name most
of them, but it is not worth while to get into a scrape
with them; they do not like people to know of their
Amager flight upon quill pens. I have a kind of a
cousin, who is a fisherman's wife, and furnishes abusive
articles to three popular periodicals : she says she has
been out there as an invited guest. She has described
the whole affair. Half that she says, of course, are lies,
but part might be true. When she was there, they commenced with a song ; each of the visitors had written his
own song, and each sang his own composition : they all
performed together, so it was a kind of ' cats' chorus.'
Small groups marched about, consisting of those who
labour at improving that gift which is called ' the gift of
the gab :' they had their own shrill songs. Then came

the little drummers, and those who write without giving their names — that is to say, whose grease is imposed on people for blacking; then there were executioners, and the puffers of bad wares. In the midst of all the merriment, as it must have been, that was going on, shot up from a pit a stem, a tree, a monstrous flower, a large toadstool, and a cupola. These were the Utopian productions of the honoured assembly, the entire amount of their offerings to the world during the past year. Sparks flew from these various objects; they were the thoughts and ideas which had been borrowed or stolen, which now took wings to themselves, and flew away as if by magic. My cousin told me a good deal more, which, though laughable, was too malicious for me to repeat.

"I always watch this wild host fly past every New Year's eve; but on the last one, as I told you, I neglected to look at them, for I was rolling away in thought upon the round pebbles — rolling through thousands and thousands of years. I saw them detached from rocks far away in the distant north; saw them driven along in masses of ice before Noah's ark was put together; saw them sink to the bottom, and rise again in a sand-bank, which grew higher and higher above the water; and I said, 'That will be Zealand!' It became the resort of birds of various species unknown to us — the home of savage chiefs as little known to us, until the axe cut the Runic characters which then brought them into our chronology. As I was thus musing three or four falling stars attracted my eye. My thoughts took another turn. Do you know what falling stars are? The scientific themselves do not know what they are. I have my own ideas

about them. How often in secret are not thanks and
blessings poured out on those who have done anything
great or good! Sometimes these thanks are voiceless,
but they do not fall to the ground. I fancy that they
are caught by the sunshine, and that the sunbeam brings
the silent, secret praise down over the head of the benefactor. If it be an entire people that through time bestow their thanks, then the thanks come as a banquet —
fall like a falling star over the grave of the benefactor.
It is one of my pleasures, especially when on a New
Year's eve I observe a falling star, to imagine to whose
grave the starry messenger of gratitude is speeding.
One of the last falling stars I saw took its blazing course
towards the south-west. For whom was it dispatched?
It fell, I thought, on the slope by Flensborg Fiord, where
the Danish flag waves over Schleppegrell's, Læssöe's, and
their comrades' graves. One fell in the centre of the
country near Sorö. It was a banquet for Holberg's
grave — a thank offering of years from many — a thank
offering for his splendid comedies! It is a glorious and
gratifying fancy that a falling star could illumine our
graves. That will not be the case with mine; not even
a single sunbeam will bring me thanks, for I have done
nothing to deserve them. I have not even attained to
blacking," said Olé; "my lot in life has been only to get
grease."

THE SECOND VISIT.

It was on a New Year's day that I again ascended to
the church tower. Olé began to speak of toasts. We

drank one to the transition from the old drop in eternity to the new drop in eternity, as he called the year. Then he gave me his story about the glasses, and there was some sense in it.

"When the clocks strike twelve on New Year's night every one rises from table with a brimful glass, and drinks to the New Year. To commence the year with a glass in one's hand is a good beginning for a drunkard. To begin the year by going to bed is a good beginning for a sluggard. Sleep will, in the course of his year, play a prominent part; so will the glass.

"Do you know what dwells in glasses?" he asked. "There dwell in them health, glee, and folly. Within them dwell, also, vexations and bitter calamity. When I count up the glasses I can tell the gradations in the glass for different people. The first glass, you see, is the glass of health; in it grow health-giving plants. Stick to that one glass, and at the end of the year you can sit peacefully in the leafy bowers of health.

"If you take the second glass a little bird will fly out of it, chirping in innocent gladness, and men will laugh and sing with it, 'Life is pleasant. Away with care, away with fear!'

"From the third glass springs forth a little winged creature — a little angel he cannot well be called, for he has Nix blood and a Nix mind. He does not come to tease, but to amuse. He places himself behind your ear, and whispers some humorous idea; he lays himself close to your heart and warms it, so that you become very merry, and fancy yourself the cleverest among a set of great wits.

"In the fourth glass is neither plant, bird, nor little figure: it is the boundary line of sense, and beyond that line let no one go.

"If you take the fifth glass you will weep over yourself — you will be foolishly happy, or become stupidly noisy. From this glass will spring Prince Carnival, flippant and crack-brained. He will entice you to accompany him; you will forget your respectability, if you have any; you will forget more than you ought or dare forget. All is pleasure, gaiety, excitement; the maskers carry you off with them; the daughters of the Evil One, in silks and flowers, come with flowing hair and voluptuous charms. Escape them if you can.

"The sixth glass! In that sits Satan himself — a well-dressed, conversable, lively, fascinating little man — who never contradicts you, allows that you are always in the right — in fact, seems quite to adopt all your opinions. He comes with a lantern to convey you home to his own habitation. There is an old legend about a saint who was to choose one of the seven mortal sins, and he chose, as he thought, the least — drunkenness; but in that state he perpetrated all the other six sins. The human nature and the devilish nature mingle. This is the sixth glass; and after that all the germs of evil thrive in us, every one of them spreading with a rapidity and vigour that cause them to be like the mustard-seed in the Bible, 'which, indeed, is the least of all seeds; but when it is grown it is the greatest among herbs, and becometh a tree.' Most of them have nothing before them but to be cast into the furnace, and be smelted there.

"This is the story of the glasses," said Olé, the

watchman of the church tower; and it applies both to those who use blacking, and to those who use only grease."

Such was the result of the second visit to Olé. More may be forthcoming at some future time.

10

Anne Lisbeth; or, The Apparition of the Beach.

ANNE LISBETH was like milk and blood, young and happy, lovely to look at; her teeth were so dazzlingly white, her eyes were so clear; her foot was light in the dance, and her head was still lighter. What did all this lead to? To no good. "The vile creature!" "She was not pretty!"

She was placed with the grave-digger's wife, and from thence she went to the count's splendid country-seat, where she lived in handsome rooms, and was dressed in silks and fineries; not a breath of wind was to blow on her; no one dared to say a rough word to her, nothing was to be done to annoy her; for she nursed the count's son and heir, who was as carefully tended as a prince, and as beautiful as an angel. How she loved that child! Her own child was away from her — he was in the grave-digger's house, where there was more hunger than plenty, and where often there was no one at home. The poor deserted child cried, but what nobody hears nobody cares about. He cried himself to sleep, and in sleep one feels neither hungry nor thirsty: sleep is, therefore, a great blessing. In the course of time Anne Lisbeth's child shot up. Ill weeds grow apace, it is said; and this poor weed grew, and seemed a member of the family who were paid for keeping him. Anne Lisbeth was quite free of him. She was a village fine lady, had everything of the best, and wore a smart bonnet whenever she went out. But she never went to the grave-digger's; it was so far from where she lived, and she had nothing to do there. The child was under their charge; *he* who paid its board could well afford it, and the child· would be taken very good care of.

The watch-dog at the lord of the manor's bleach-field sits proudly in the sunshine outside of his kennel, and growls at every one that goes past. In rainy weather he creeps inside, and lies down dry and sheltered. Anne Lisbeth's boy sat on the side of a ditch in the sunshine, amusing himself by cutting a bit of stick. In spring he saw three strawberry bushes in bloom: they would surely bear fruit. This was his pleasantest thought; but there was no fruit. He sat out in the drizzling rain, and in the heavy rain — was wet to the skin — and the sharp wind dried his clothes upon him. If he went to the farm-houses near, he was thumped and shoved about. He was "grim-looking and ugly," the girls and the boys said. What became of Anne Lisbeth's boy? What *could* become of him? It was his fate to be "*never loved.*"

At length he was transferred from his joyless village life to the still worse life of a sailor boy. He went on board a wretched little vessel, to stand by the rudder while the skipper drank. Filthy and disgusting the poor boy looked; starving and benumbed with cold he was. One would have thought, from his appearance, that he never had been well fed; and, indeed, that was the fact.

It was late in the year; it was raw, wet, stormy weather; the cold wind penetrated even through thick clothing, especially at sea; and only two men on board were too few to work the sails; indeed, it might be said only one man and a half — the master and his boy. It had been black and gloomy all day; now it became still more dark, and it was bitterly cold. The skipper took a dram to warm himself. The flask was old, and so was the glass; its foot was broken off, but it was inserted into a piece of wood painted blue, which served as a stand for it. If one dram was good, two would be better, thought the master. The boy stood by the helm, and held on to it with his hard, tar-covered hands. He looked frightened. His hair was rough, and he was wrinkled, and stunted in his growth. The young sailor was the grave-digger's boy; in the church register he was called Anne Lisbeth's son.

The wind blew as it list; the sail flapped, then filled; the vessel flew on. It was wet, chill, dark as pitch; but worse was yet to come. Hark! What was that? With what had the boat come in contact? What had burst? What seemed to have caught it? It shifted round. Was it a sudden squall? The boy at the helm cried aloud, " In the name of Jesus !" The little bark had struck on a large sunken rock, and sank as an old shoe would sink in a small pool — sank with men and mice on board, as the saying is; and there certainly were mice, but only one man and a half — the skipper and the grave-digger's boy.

None witnessed the catastrophe except the screaming sea-
gulls and the fishes below; and even they did not see
much of it, for they rushed aside in alarm when the water gushed thundering into the little vessel as it sank.
Scarcely a fathom beneath the surface it stood; yet the
two human beings who had been on board were lost —
lost — forgotten! Only the glass with the blue-painted
wooden foot did not sink; the wooden foot floated it.
But the glass was broken when it was washed far up on
the beach. How and when? That is of no consequence.
It had served its time, and it had been liked; that Anne
Lisbeth's child had never been. But in the kingdom of
heaven no soul can say again, " Never loved! "

Anne Lisbeth resided in the large market town, and
had done so for some years. She was called " Madam,"
and held her head very high, especially when she spoke
of old reminiscences of the time she had passed at the
count's lordly mansion, when she used to drive out in a
carriage, and used to converse with countesses and baron-
esses. Her sweet nursling, the little count, was a lovely
angel, a darling creature. She was so fond of him, and
he had been so fond of her. How she used to pet him,
and how he used to kiss her! He was her delight —
was as dear to her as herself. He was now quite a big
boy; he was fourteen years of age, and had plenty of
learning and accomplishments. She had not seen him
since she carried him in her arms. It was many years
since she had been at the count's castle, for it was such a
long way off.

" But I must go over and see them again," said Anne

Lisbeth. "I must go to my noble friends, to my darling
child, the young count — yes, yes, for he is surely longing to see me. He thinks of me, he loves me as he did
when he used to throw his little cherub arms round my
neck and lisp, 'An Lis!' Oh, it was like a violin! Yes,
I must go over and see him again."

She went part of the way in the carrier's wagon, part
of the way on foot. She arrived at the castle. It looked
as grand and imposing as ever. The gardens were not
at all changed; but the servants were all strangers.
Not one of them knew anything about Anne Lisbeth.
They did not know what an important person she had
been in the house formerly; but surely the countess
would tell them who she was, so would her own boy.
How she longed to see them both!

Well, Anne Lisbeth was there; but she had to wait a
long time, and waiting is always so tedious. Before the
family and their guests went to dinner, she was called in
to the countess, and very kindly spoken to. She was
told she should see her dear boy after dinner, and after
dinner she was sent for again.

How much he had grown! How tall and thin! But
he had the same charming eyes, and the same angelic
mouth. He looked at her, but he did not say a word.
It was evident that he did not remember her. He
turned away, and was going, but she caught his hand and
carried it to her lips. "Ah! well, that will do!" he
said, and hastily left the room — he, the darling of her
soul — he on whom her thoughts had centred for so
many years — he whom she had loved the best — her
greatest earthly pride!

Anne Lisbeth left the castle, and turned into the open high road. She was very sad — he had been so cold and distant to her. He had not a word, not a thought for her who, by day and by night, had so cherished *him* in her heart.

At that moment a large black raven flew across the road before her, screeching harshly.

"Oh!" she exclaimed, "what do you want, bird of ill omen that you are?"

She passed by the grave-digger's house: his wife was standing in the doorway, and they spoke to each other.

"You are looking very well," said the grave-digger's wife. "You are stout and hearty. The world goes well with you apparently."

"Pretty well," replied Anne Lisbeth.

"The little vessel has been lost," said the grave-digger's wife. "Lars the skipper, and the boy, are both drowned; so there is an end of that matter. I had hoped, though, that the boy might by and by have helped me with a shilling now and then. He never cost you anything, Anne Lisbeth."

"Drowned are they?" exclaimed Anne Lisbeth; and she did not say another word on the subject — she was so distressed that her nursling, the young count, did not care to speak to her — she who loved him so much, and had taken such a long journey to see him — a journey that had cost her some money too. The pleasure she had received was not great, but she was not going to admit this. She would not say one word to the grave-digger's wife to lead her to think that she was no longer a person of consequence at the count's. The raven screeched again just over her head.

"That horrid noise!" said Anne Lisbeth; "it has quite startled me to-day."

She had brought some coffee-beans and chicory with her; it would be a kindness to the grave-digger's wife to make her a present of these; and, when she did so, it was agreed that they should take a cup of coffee together. The mistress of the house went to prepare it, and Anne Lisbeth sat down to wait for it. While waiting she fell asleep, and she dreamed of one of whom she had never before dreamt: that was very strange. She dreamed of her own child, who in that very house had starved and squalled, and never tasted anything better than cold water, and who now lay in the deep sea, our Lord only knew where. She dreamed that she was sitting just where she really was seated, and that the grave-digger's wife had gone to make some coffee, but had first to grind the coffee-beans, and that a beautiful boy stood in the doorway — a boy as charming as the little count had been ; and the child said, —

"The world is now passing away. Hold fast to me, for thou art my mother. Thy child is an angel in the kingdom of heaven. Hold fast to me!"

And he seized her. But there was a frightful uproar around, as if worlds were breaking asunder; and the angel raised her up, and held her fast by the sleeves of her dress — so fast, it seemed to her, that she was lifted from the ground; but something hung so heavily about her feet, something lay so heavily on her back: it was as if hundreds of women were clinging fast to her, and crying, "If thou canst be saved, so may we. We will hold on — hold on!" and they all appeared to be holding on by her. Then the sleeves of her garments gave way, and she fell, overcome with terror.

The sensation of fear awoke her, and she found herself on the point of falling off her chair. Her head was so confused that at first she could not remember what she had dreamt, though she knew it had been something disagreeable. The coffee was drunk, and Anne Lisbeth took her departure to the nearest village, where she might meet the carrier, and get him to convey her that evening to the town where she lived. But the carrier said he was not going until the following evening; and, on calculating what it would cost her to remain till then, she determined to walk home. She would not go by the high road, but by the beach: that was at least eight or nine miles shorter. The weather was fine, and it was full moon. She would be at home the next morning.

The sun had set; the evening bells that had been chiming were hushed. All was still; not a bird was to be heard twittering among the leaves — they had all gone to rest: the owls were away. All was silence in the wood; and on the beach, where she was walking, she could hear her own foot fall on the sand. The very sea seemed slumbering; the waves rolled lazily and noiselessly on the shore, and away on the open deep there seemed to be a dead calm: not a line of foam, not a ripple was visible on the water. All were quiet beneath, the living and the dead.

Anne Lisbeth walked on, and her thoughts were not engrossed by anything in particular. She was not at all lost in thought, but thoughts were not lost to her. They

10 *   o

are never lost to us; they lie only in a state of torpor, as it were, both the lately active thoughts that have lulled themselves to rest, and those which have not yet awoke. But thoughts come often undesired; they can touch the heart, they can distract the head, they can at times over-power us.

"Good actions have their reward," it is written.

"The wages of sin is death," it is also written. Much is written — much is said. But many give no heed to the words of truth — they remember them not; and so it was with Anne Lisbeth; but they can force themselves upon the mind.

All sins and all virtues lie in our hearts — in thine, in mine. They lie like small invisible seeds. From with-out fall upon them a sunbeam, or the contact of an evil hand—they take their bent in their hidden nook, to the right or to the left. Yes, there it is decided, and the little grain of seed quivers, swells, springs up, and pours its juice into your blood, and there you are, fairly launched. These are thoughts fraught with anxiety; they do not haunt one when one is in a state of mental slumber, but they are fermenting. Anne Lisbeth was slumbering — hidden thoughts were fermenting. From Candlemas to Candlemas the heart has much on its tablets—it has the year's account. Much is forgotten — sins in word and deed against God, against our neighbour, and against our own consciences. We reflect little upon all this; neither did Anne Lisbeth. She had not broken the laws of her country, she kept up good appearances, she did not run in debt, she wronged no one; and so, well satisfied with herself, she walked on by the seashore. What was that lying in her path? She stopped. What was that washed up from the sea? A man's old hat lay there. It might have fallen overboard. She approached closer to it, stood still, and looked at it. Heavens! what was lying there? She was almost frightened; but there was nothing to be frightened at; it was only a mass of seaweed that lay twined over a large, oblong, flat rock, that was shaped something like a human being — it was nothing but seaweed. Still she felt frightened, and hastened on; and as she hurried on, many things she had heard in her childhood recurred to her thoughts, especially all the superstitious tales about *" the apparition of the beach "* — the spectre of the unburied that lay washed up on the lonely, deserted shore. The body thrown up from the deep, the dead body itself, she thought nothing of; but its ghost followed the solitary wanderer, attached itself closely to her, and demanded to be carried to the churchyard, to receive Christian burial.

" Hold on — hold on!" it was wont to say; and, as Anne Lisbeth repeated these words inwardly to herself, she suddenly remembered her strange dream, in which the women had clung to her, shrieking, " Hold on — hold on!" how the world had sunk; how her sleeves had given way, and she had fallen from the grasp of her child, who wished, in the hour of doom, to save her. Her child — her own flesh and blood — the little one she had never loved, never spared a thought to — that child was now at the bottom of the sea, and it might come like " the apparition of the beach," and cry, " Hold on — hold on! Give me Christian burial!" And as these thoughts crowded on her mind, terror gave wings to her feet, and she hurried faster and faster on; but fear came like a cold, clammy hand, and laid itself on her beating heart, so that she felt quite faint; and as she glanced towards the sea, she saw it looked dark and threatening; a thick mist arose, and soon spread around, lying heavily over the very trees and bushes, which assumed strange appearances through it.

She turned round to look for the moon, which was behind her: it was like a pale disc, without any rays. Something seemed to hang heavily about her limbs as she attempted to hurry on. She thought of the apparition; and, turning again, she beheld the white moon as if close to her, while the mist seemed to hang like a mantle over her shoulders. " Hold on — hold on! Give me Christian burial!" she expected every moment to hear; and she did hear a hollow, terrific sound, which seemed to cry hoarsely, " Bury me — bury me!" Yes, it must be the spectre of her child — her child who was lying at the bottom of the sea, and who would not rest quietly until the corpse was carried to the churchyard, and placed like a Christian in consecrated ground. She would go there — she would dig his grave herself; and she went in the direction in which the church lay, and as she proceeded she felt her invisible burden become lighter — it left her; and again she returned to the shore to reach her home as speedily as possible. But no sooner did her foot tread the sands than the wild sound seemed to moan around her, and it seemed ever to repeat, " Bury me — bury me!"

The fog was cold and damp; her hands and her face

were cold and damp. She shivered in her fright. Without, space seemed to close up around her; within her
there seemed to be endless room for thoughts that had
never before entered her mind.

During one spring night here in the north the beech
groves can sprout, and the next day's early sun can shine
on them in all their fresh young beauty. In one single
second within us can the germ of sin bud forth, swelling
by degrees into thoughts, words, and deeds, though all
remorse for them lies dormant. *It* is quickened and
unfolds itself in one single second, when conscience
awakens; and our Lord awakens *that* when we least
expect it. Then there is nothing to be excused; deeds
stand forth and bear witness, thoughts find words, and
words ring out over the world. We are shocked at what
we have permitted to dwell within us, and not stifled;
shocked at what, in our thoughtlessness or our presump-
tion, we have scattered abroad. The heart is the depos-
itory of all virtues, but also of all vices; and these can
thrive in the most barren ground.

Anne Lisbeth reviewed in thought what we have expressed in words. She was overwhelmed with it all.
She sank to the ground, and crawled a little way over it.
" Bury me — bury me ! " she still seemed to hear. She
would rather have buried herself, if the grave could be
an eternal forgetfulness of everything. It was the awakening hour of serious thought, of terrible thoughts, that
made her shudder. Superstition came, too, by turns
heating and chilling her blood; and things she would
scarcely have ventured to mention rushed on her mind.
Noiseless as the clouds that crossed the sky in the clear

moonlight floated past her a vision she had heard of.
Immediately before her sped four foaming horses, flames
flashing from their eyes and from their distended nostrils;
they drew a fiery chariot, in which sat the evil lord of
the manor, who, more than a hundred years before, had
dwelt in that neighbourhood. Every night, it is said, he
drives to his former home, and then instantly turns back
again. He was not white, as the dead are said to be;
no, he was as black as a coal — a burnt-out coal. He
nodded to Anne Lisbeth, and beckoned to her: " Hold
on — hold on! So mayst thou again drive in a nobleman's carriage, and forget thine own child ! "

In still greater terror, and with still greater precipita-
tion than before, she fled in the direction of the church.
She reached the churchyard ; but the dark crosses above
the graves, and the dark ravens, seemed to mingle together before her eyes. The ravens screeched as they
had screeched in the daytime ; but she now understood
what they said, and each cried, " I am a raven-mother ;
I am a raven-mother !." And Anne Lisbeth thought
that they were taunting her. She fancied that she
might, perhaps, be changed into such a dark bird, and
might have to screech like them, if she could not get the
grave demanded of her dug.

And she threw herself down upon the ground, and she
dug a grave with her hands in the hard earth, so that
blood sprang from her fingers.

" Bury me — bury me ! " resounded still about her.
She dreaded the crowing of the cock, and the first red
streak in the east, because, if they came before her labours were ended, she would be lost. And the cock

crowed, and in the east it began to be light. The grave was but half dug. An ice-cold hand glided over her head and her face, down to where her heart was. "Only half a grave!" sighed a voice near her; and something seemed to vanish away — vanish into the deep sea. It was "the apparition of the beach." Anne Lisbeth sank, terror-stricken and benumbed, on the ground. She had lost feeling and consciousness.

It was broad daylight when she came to herself. Two young men lifted her up. She was lying, not in the churchyard, but down on the shore; and she had dug there a deep hole in the sand, and cut her fingers till they bled with a broken glass, the stem of which was stuck into a piece of wood painted blue. Anne Lisbeth was ill. Conscience had mingled in Superstition's game, and had imbued her with the idea that she had only half a soul — that her child had taken the other half away with him down to the bottom of the sea. Never could she ascend upwards towards the mercy-seat, until she had again the half soul that was imprisoned in the depths of the ocean. Anne Lisbeth was taken to her home, but she never was the same as she had formerly been. Her thoughts were disordered like tangled yarn; one thread alone was straight — that was to let "the apparition of the beach" see that a grave was dug for him in the churchyard, and thus to win back her entire soul.

Many a night she was missed from her home, and she was always found on the seashore, where she waited for the spectre of the dead. Thus passed a whole year. Then she disappeared one night, and was not to be found. The whole of the next day they searched for her in vain.

Towards the evening, when the bell-ringer entered
the church to ring the evening chimes, he saw Anne
Lisbeth lying before the altar. She had been there
from a very early hour in the morning; her strength
was almost exhausted, but her eyes sparkled, her face
glowed with a sort of rosy tint. The departing rays
of the sun shone in on her, and streamed over the
altar-piece, and on the silver clasps of the Bible, that
lay open at the words of the prophet Joel: "Rend
your heart, and not your garments, and turn unto the
Lord your God." "It was a strange occurrence," people said — as if everything were chance.

On Anne Lisbeth's countenance, when lighted up
by the sun, were to be read peace and comfort. "She
felt so well," she said. "She had won back her soul."
During the night "the apparition of the beach" —
her own child — had been with her, and it had said, —

"Thou hast only dug half a grave for me; but now
for a year and a day thou hast entombed me in thy
heart, and there a mother best inters her child." And
he had restored to her her lost half soul, and had led
her into the church.

"Now I am in God's house," said she, "and in it
one is blessed."

When the sun had sunk entirely Anne Lisbeth's spirit
had soared far away up yonder, where there is no more
fear when one's sins are blotted out; and hers, it might
be hoped, had been blotted out by the Saviour of the
world.

Children's Prattle.

A T the merchant's house there was a large party of children — rich people's children and great people's children. The merchant was a man of good standing in society, and a learned man. He had taken, in his youth, a college examination. He had been kept to his studies by his worthy father, who had not gone very deep into learning himself, but was honest and active. He had made money, and the merchant had increased the fortune left to him. He had intellect, and heart too; but less was said of these good qualities than of his money.

There visited at his house several distinguished persons, both people of birth, as it is called, and people of talents, as it is called — people who came under both of these heads, and people who came under neither of these heads. The meeting now in question was a children's party, where there was childish talk; and children generally speak like parrots.

There was one little girl so excessively proud. She had been flattered into her foolish pride by the servants, not by her parents — they were too sensible to have done that. Her father was *Kammerjunker* * and she thought this was monstrously grand.

"I am a court child," she said.

She might as well have been a cellar child, as far as she was herself concerned; and she informed the other children that she was "born" (*well born*, she meant); that when people were not "born," they could never be anybody; and that, however much they might read, however clever and industrious they might be, if they were not "born" they could never become great.

"And those whose names end in '*sen*,'" she continued, "are all low people, and can never be of any consequence in the world. Ladies and gentlemen would put their hands on their sides, and keep them at a distance, these 'sen — sens'!" And she threw herself into the attitude she had described, and stuck her pretty little arms akimbo, to show how people of her grade would carry themselves in the presence of such common creatures. She really looked very pretty.

But the merchant's little daughter became extremely angry. Her father was called "Madsen," and that name, she knew, ended in "sen;" so she said, as proudly as she could, —

"But my father can buy hundreds of rix dollars' worth of sugar-plums, and think nothing of it. Can your father do that?"

"That's all very well," said the little daughter of a popular journalist; "but my father can put both of your fathers and all 'fathers' into the newspaper.

---

* A title at court.

Every one is afraid of him, my mother says; for it is my father who rules everything through the newspaper." And the little girl tossed her head and strutted about as if she thought herself a princess.

But on the outside of the half-open door stood a poor little boy peeping in. It was, of course, out of the question that so poor a child should enter the drawing-room; but he had been turning the spit for the cook, and he had obtained permission to look in behind the door at the splendidly dressed children who were amusing themselves, and that was a treat to him.

He would have liked to have been one of them, he thought; but at that moment he heard what had been said, and it was enough to make him very sad. Not one shilling had his parents at home to spare. They were not able to set up a newspaper, to say nothing of writing for one. And the worse was yet to come; for his father's name, and of course also his own name, certainly ended in "sen." He, therefore, could never become anybody in this world. This was very disheartening. Though he felt assured that he was *born*, it was impossible to think otherwise.

This was what passed that evening.

Several years had elapsed, and during their course the children had grown up to be men and women.

There stood in the town a handsome house, which was filled with magnificent objects of art. Every one went to see it. Even people who lived at a distance came to town to see it. Which prodigy, among the children we have spoken of, could call that edifice his or hers? It is easy to tell that. No; it is not so easy, after all. That house belonged to the poor little boy, who became somebody, although his name *did* end in "sen." — THORWALDSEN!

And the three other children — the children of high birth, money, and literary arrogance? Yes; there is nothing to be said about them. They are all alike. They grew up to be all very respectable, comfortable, and commonplace. They were well-meaning people. What they had formerly said and thought was only — CHILDREN'S PRATTLE.

A Row of Pearls.

I.

THE railroad in Denmark extends no farther as yet than from Copenhagen to Korsör. It is a row of pearls. Europe has a wealth of these. Its most costly pearls are named Paris, London, Vienna, Naples; though many a one does not point out these great cities as his most beautiful pearl, but, on the contrary, names some small, by no means remarkable town, for it is *his* home — the home where those he loves reside. Nay, sometimes it is but a country seat — a small cottage hidden among green hedges — a mere spot that he hastens towards, while the railway train rushes on.

How many pearls are there upon the line from Copenhagen to Korsör? We will say six. Most people must remark these. Old remembrances and poetry itself bestow a radiance on these pearls, so that they shine in on our thoughts.

Near the rising ground where the palace of Frederick VI. stands — the home of Ochlenschläger's childhood — shines, under the lee of Sondermarken's woody ground, one of these pearls. It is called the " Cottage of Phile-

mon and Baucis;" that is to say, the home of two loving
old people. Here dwelt Rahbek and his wife Camma ;
here, under their hospitable roof, were collected from the
busy Copenhagen all the superior intellects of their day ;
here was the home of genius ; and now say not, "Ah,
how changed!"   No; it is still the spirits' home — a
hothouse for sickly plants.   Buds that are not strong
enough to expand into flowers, preserve, though hidden,
all the germs of a luxuriant tree.   Here the sun of mind
shines in on a home of stagnant spirits, reviving and
cheering it.   The world around beams through the eyes
into the soul's unfathomable depths.   *The Idiot's Home*,
surrounded by the love and kindness of human beings, is
a holy place — a hothouse for those sickly plants that
shall in future be transplanted to bloom in the garden of
paradise.   The weakest in the world are now gathered
here, where once the greatest and the wisest met, ex-
changed thoughts, and were lifted upwards.   Their mem-
ories will ever be associated with the " Cottage of Phile-
mon and Baucis."

The burial-place of kings by Hroar's spring — the
ancient Roeskilde — lies before us.   The cathedral's
slender spires tower over the low town, and are reflected
on the surface of the fiord.   One grave alone shall we
seek here ; that shall not be the tomb of the mighty
Margrethe — the union queen.   No; within the churchyard, near whose white walls we have so closely flown,
is the grave: a humble stone is laid over it.   Here
reposes the great organist — the reviver of the old
Danish romances.   With the melodies we can recall the
words, —

"The clear waves rolled,"

and

"There dwelt a king in Leiré." *

Roeskilde! thou burial-place of kings, in thy pearl we shall see the lonely grave on whose stone is chiselled a lyre and the name — WEYSE.

Now come we to Sigersted, near Ringsted. The river is shallow — the yellow corn waves where Hagbarth's boat was moored, not far from Signé's maiden bower. Who does not know the tradition about Hagbarth † and Signelil, and their passionate love — that Hagbarth was hanged in the galley, while Signelil's tower stood in flames ?

---

* Leiré, the original residence of the Danish kings, said to have been founded by Skiold, a son of Odin, was, during the heathen ages, a place of note. It contained a large and celebrated temple for offerings, to which people thronged every ninth year, at the period of the great Yule feast, which was held annually in mid-winter, commencing on the 4th of January. In Norway this ancient festival was held in honour of Thor; in Denmark, in honour of Odin. Every ninth year the sacrifices were on a larger scale than usual, consisting then of ninety-nine horses, dogs, and cocks — human beings were also sometimes offered. When Christianity was established in Denmark, the seat of royalty was transferred to Roeskilde, and Leiré fell into total insignificance. It is now merely a village in Zealand. — *Trans.*

† Hagbarth, a son of the Norwegian king, Amund, and his three brothers, Hake, Helvin, and Hamund, scoured the seas with a hundred ships, and fell in with the king of Zealand's three sons, Sivald, Alf, and Alger. They attacked each other, and continued their bloody strife until a late hour at night. Next day they all found their ships so disabled that they could not renew the conflict. Thereupon they made friends, and the Norwegian princes or pirates accompanied the Zealanders to the court of their father, King Sigar. Here Hagbarth won the heart of the king's daughter Signé, and they became secretly engaged. Hildigeslev, a handsome German prince, was at that time

"Beautiful Sorö, encircled by woods!" thy tranquil,
cloistered town peeps forth from among thy moss-covered
trees; the keen bright eyes of youth gaze from the
academy, over the lake, to the busy highway, where the
locomotive's dragon snorts, while it is flying through the
wood. Sorö, thou poet's pearl, that hast in thy custody
the honoured dust of Holberg! like a majestic white
swan by the deep lake stands thy far-famed seat of
learning. We fix our eyes on it, and then they wander
in search of the simple star-flower in the wooded ground
— a small house. Pious hymns are chanted there, that
echo over the length and breadth of the land; words are
uttered there to which the very rustics listen, and hear of

---

her suitor; but she refused him, and in revenge he sowed discord
between her lover and his brothers and her brothers. Alf and Alger
murdered Hagbarth's brothers, Helvin and Hamund, but were killed
in their turn by Hagbarth and Hake. After this deed Hagbarth dared
not remain at Sigar's court; but he longed so much to be with Signé,
that he dressed himself as a woman, and in this disguise he obtained
admission to the palace, and contrived to be named one of her attendants. The damsels of her suite were much surprised at the hardness
of the new waiting-maid's hands, and at other unfeminine peculiarities
which they remarked; but Signé appointed him her especial attend-
ant, and thus partially removed him from their troublesome curiosity.
Fancying themselves safe, they relaxed their precautions. Hagbarth
was discovered, secured, and carried before the *Thing*, or judicial
assembly. Before he left her he received a promise from Signé that
she would not survive him. He was condemned to death; to be
hanged on board a galley, in view of Signé's dwelling. To prove her
love and faith, he entreated that his mantle might be hung up first, in
order, he said, that the sight of it might prepare him for his own
death. It was done; and when Signé saw it, she fancied her lover
was dead, and instantly set fire to her abode. Hagbarth beheld the
flames; and no longer doubting the constancy of the princess, he died
rejoicing in following her to the other world. — *Trans.*

Denmark's bygone ages. As the greenwood and the birds' songs belong to each other, so are associated the names of Sorö and INGEMANN.

To Slagelsé! What is the pearl that dazzles us here? The monastery of Antoorskov has vanished, even the last solitary remaining wing, though one old relic still exists — renovated and renovated again — a wooden cross upon the heights above, where, in legendary lore, it is said that HOLY ANDERS, the warrior priest, woke up, borne thither in one night from Jerusalem!

Korsör — there wert thou * born, who gave us

> "Mirth with melancholy mingled,
> In stories of ' Knud Sjællandsfar.'

Thou master of language and of wit! the old decaying ramparts of the deserted fortification are now the last visible mementos of thy childhood's home. When the sun is sinking, their shadows fall upon the spot where stood the house in which thine eyes first opened on the light. From these ramparts, looking towards Sprogōs hills, thou sawest, when thou "wert little,"

> "The moon behind the island sink;"

and sang it in undying verse, as afterwards thou didst sing the mountains of Switzerland; thou, who didst wander through the vast labyrinth of the world, and found that

> "Nowhere do the roses seem so red —
> Ah! nowhere else the thorn so small appears,
> And nowhere makes the down so soft a bed,
> As that where innocence reposed in bygone years!"

---

* Jeus Baggesen. — *Trans.*

11   P

Capricious, charming warbler! We will weave a wreath of woodbine. We will cast it into the waves, and they will bear it to Kielerfiord, upon whose coast thine ashes repose. It will bring a greeting from a younger race, a greeting from thy native town, Korsōr, where ends the row of pearls.

II.

"It is, truly enough, a row of pearls from Copenhagen to Korsōr," said my grandmother, who had heard read· aloud what we have just been reading. "It is a row of pearls for me, and it was that more than forty years ago," she added. "We had no steam engines then. It took us days to make a journey which you can make now in a few hours. For instance, in 1815, I was then one-and-twenty years old. That is a pleasant age. Even up in the thirties it is also a pleasant age. In my young days it was much rarer than now to go to Copenhagen, the city of all cities, as we thought it. After twenty years' absence from it, my parents determined to visit it once more, and I was to accompany them. The journey had been projected and talked of for years. At length it was positively to be accomplished. I fancied that I was beginning quite a new life, and certainly, in one way, a new life did begin for me.

"After a great deal of packing and preparations we were ready to start. Then what numbers of our neighbours came to bid us good-by! It was a very long journey we had before us. Shortly before mid-day we drove out of Odense in my father's Holstern wagon — a roomy carriage. Our acquaintances bowed to us from the windows of almost every house until we were outside of St. Jorgen's Port. The weather was delightful, the birds were singing, all was pleasure. We forgot that it was a long way and a rough road to Nyborg. We reached that place towards evening. The post did not arrive till midnight, and until it came the packet could not sail. At length we went on board. Before us lay the wide waters, as far as the eye could see, and it was a dead calm. We lay down in our clothes and slept. When I awoke in the morning, and went on deck, nothing could be seen on either side of us, there was such a thick fog. I heard the cocks crowing, and I knew the sun must have risen. Bells were ringing : where could they be? The mist cleared away, and we found we were lying a litle way from Nyborg. As the day advanced we had a little wind : it stiffened, and we got on faster. At last we were so fortunate, at a little after eleven o'clock at night, as to reach Korsōr. We had taken twenty-two hours to go sixteen miles.

" Glad we were to land; but it was extremely dark, and the lanterns gave very little light. However, all was wonderful to me, who had never been in any other town but Odense.

" ' Here Baggesen was born,' said my father, ' and here Birckner lived.'

" It seemed to me that the old town, with its small houses, became at once larger and more important. We were also rejoiced to have the firm earth under us once more ; but I could not sleep that night, I was so excited thinking over all I had seen and encountered since I had left home two days before.

" Next morning we rose early. We had before us a
bad road, with frightful hills and many valleys, till we
reached Slagelsé; and beyond it, on the other side, it
was but little better; therefore we were anxious to get
to Krebsehuset, that we might early next day go on to
Sorö, and visit Möllers Emil, as we called him. He was
your grandfather, my worthy husband, the dean. He
was then a student at Sorö, and very busy about his
second examination.

" Well, we arrived about noon at Krebsehuset. It was
a gay little town then, and had the best inn on the road,
and the prettiest country round it: you must all admit
that it is pretty still. She was a very active landlady,
Madame Plambek, and everything in her house was as
clean as a new pin. There hung up on her wall a letter
from Baggesen to her. It was framed, and had a glass
over it; it was a very interesting object to look at, and to
me it was quite a curiosity. We then went into Sorö, and
found Emil there. You may believe he was very glad
to see us, and we were very glad to see him — he was
so good and so attentive. We went with him to see the
church, with Absolon's grave and Holberg's coffin. We
saw the old monkish inscriptions, and we sailed over the
lake to Parnasset — the sweetest evening I remember.
I recollect well that I thought, if any one could write
poetry anywhere in the world, it would be at Sorö, amidst
those charming, peaceful scenes, where nature reigns in
all her beauty. Afterwards we visited by moonlight the
' Philosopher's Walk,' as it was called — the beautiful,
lonely path by the lake and the moor that leads towards
the highway to Krebsehuset. Emil remained to supper

with us, and my father and mother thought he had become very clever and very good-looking. He promised us that he would be in Copenhagen within a few days, and would join us there: it was then Whitsuntide. We were going to stay with his family. These hours at Sorö and Krebsehuset, may they not be deemed the most beautiful pearls of my life?

"The next morning we commenced our journey at a very early hour, for we had a long way to go to reach Roeskilde, and we were anxious to get there in time to see the church. In the evening my father wished to visit an old friend, so we stopped at Roeskilde that night, and the next day we arrived at Copenhagen. It took us three days to go from Korsōr to Copenhagen; now the journey is made in three hours. The pearls have not become more valuable — that they could not be — but they are strung together in a new and wonderful manner. I remained three weeks with my parents in Copenhagen, and Emil was with us there for a fortnight. When we returned to Fyen, he accompanied us as far as Korsōr. There, before parting, we were betrothed; so you can well believe that *I* call from Copenhagen to Korsōr a row of pearls.

"Afterwards, when Emil and I were married, we often spoke of the journey to Copenhagen, and of undertaking it once more. But then came first your mother, then she had brothers and sisters, and there was a great deal to do; so the journey was put off. And when your grandfather got preferment, and was made dean, all was thankfulness and joy; but we never got to Copenhagen. No, never have I set foot in it again, as often as we thought of it and projected going. Now I am too old, and I could not stand travelling by a railroad; but I am very glad there are railroads — they are a blessing to many. You can come more speedily to me; and Odense is now not farther from Copenhagen than in my younger days it was from Nyborg. You could now go in almost the same space of time to Italy as it took us to travel to Copenhagen. Yes, that is something!

"Nevertheless, I shall stay in one place, and let others travel and come to me if they please. But you should not laugh at me for keeping so quiet; I have a greater journey before me than any by the railroad. When it shall please our Lord, I have to travel up to your grandfather; and when you have finished your appointed time on earth, and enjoyed the blessings bestowed here by the Almighty, then I trust that you will ascend to us; and if we then revert to our earthly days, believe me, children, I shall say then as now, 'From Copenhagen to Korsör is indeed 'A ROW OF PEARLS.'"

The Pen and the Inkstand.

THE following remark was made in a poet's room, as the speaker looked at the inkstand that stood upon his table : —

"It is astonishing all that can come out of that inkstand What will it produce next? Yes, it is wonderful!"

"So it is!" exclaimed the inkstand. "It is incomprehensible! That is what I always say." It was thus the inkstand addressed itself to the pen, and to everything else that could hear it on the table. "It is really astonishing all that can come from me! It is almost incredible! I positively do not know myself what the next production may be, when a person begins to dip into me. One drop of me serves for half a side of paper; and what may not then appear upon it? I am certainly something extraordinary. From me proceed all the works of the poets. These animated beings, whom people think they recognise — these deep feelings, that gay humour, these charming descriptions of nature — I do not understand them myself, for I know nothing about nature; but still it is all in me. From me have gone

forth, and still go forth, these warrior hosts, these lovely
maidens, these bold knights on snorting steeds, those droll
characters in humbler life. The fact is, however, that I
do not know anything about them myself. I assure you
they are not my ideas."

"You are right there," replied the pen. "You have
few ideas, and do not trouble yourself much with thinking. If you *did* exert yourself to think, you would perceive that you ought to give something that was not dry.
You supply me with the means of committing to paper
what I have in me; I write with that. It is the pen that
writes. Mankind do not doubt that; and most men have
about as much genius for poetry as an old inkstand."

"You have but little experience," said the inkstand.
"You have scarcely been a week in use, and you are
already half worn out. Do you fancy that you are a
poet? You are only a servant; and I have had many
of your kind before you came — many of the goose family, and of English manufacture. I know both quill pens
and steel pens. I have had a great many in my service,
and I shall have many more still, when he, the man who
stirs me up, comes and puts down what he takes from
me. I should like very much to know what will be the
next thing he will take from me."

Late in the evening the poet returned home. He had
been at a concert, had heard a celebrated violin player,
and was quite enchanted with his wonderful performance.
It had been a complete gush of melody that he had drawn
from the instrument. Sometimes it seemed like the gentle murmur of a rippling stream, sometimes like the singing of birds, sometimes like the tempest sweeping through

the mighty pine forests. He fancied he heard his own heart weep, but in the sweet tones that can be heard in a woman's charming voice. It seemed as if not only the strings of the violin made music, but its bridge, its pegs, and its sounding-board. It was astonishing! The piece had been a most difficult one; but it seemed like play — as if the bow were but wandering capriciously over the strings. Such was the appearance of facility, that every one might have supposed he could do it. The violin seemed to sound of itself, the bow to play of itself. These two seemed to do it all. One forgot the master who guided them, who gave them life and soul. Yes, they forgot the master: but the poet thought of him. He named him, and wrote down his thoughts as follows : —

" How foolish it would be of the violin and the bow, were they to be vain of their performance! And yet this is what so often we of the human species are. Poets, artists, those who make discoveries in science, military and naval commanders — we are all proud of ourselves ; and yet we are all only the instruments in our Lord's hands. To Him alone be the glory! We have nothing to arrogate to ourselves."

This was what the poet wrote; and he headed it with, " The Master and the Instruments." When the inkstand and the pen were again alone, the latter said, —

" Well, madam, you heard him read aloud what I had written."

" Yes, what I gave you to write," said the inkstand. " It was a hit at you for your conceit. Strange that you cannot see that people make a fool of you! I gave you

11*

that hit pretty cleverly.   I confess, though, it was rather malicious."

"Ink-holder!" cried the pen.

"Writing-stick!" cried the inkstand.

They both felt assured that they had answered well; and it is a pleasant reflection that one has made a smart reply — one sleeps comfortably after it.   And they both went to sleep; but the poet could not sleep.   His thoughts welled forth like the tones from the violin, murmuring like a pearly rivulet, rushing like a storm through the forest.   He recognised the feelings of his own heart — he perceived the gleam from the everlasting Master.

To Him alone be the glory!

The Child in the Grave.

THERE was sorrow in the house, there was sorrow in the heart; for the youngest child, a little boy of four years of age, the only son, his parents' present joy and future hope, was dead. Two daughters they had, indeed, older than their boy — the eldest was almost old enough to be confirmed — amiable, sweet girls they both were; but the lost child is always the dearest, and he was the youngest, and a son. It was a heavy trial. The sisters sorrowed as young hearts sorrow, and were much afflicted by their parents' grief; the father was weighed down by the affliction; but the mother was quite overwhelmed by the terrible blow. By night and by day had she devoted herself to her sick child, watched by him, lifted him, carried him about, done everything for him herself. She had felt as if he were a part of herself: she could not bring herself to believe that he was dead — that he should be laid in a coffin, and concealed in the grave. God would not take that child from her — O no! And when he was taken, and she could no longer refuse to believe the truth, she exclaimed in her wild grief, —

" God has not ordained this! He has heartless agents here on earth. They do what they list — they hearken not to a mother's prayers !"

She dared in her woe to arraign the Most High; and then came dark thoughts, the thoughts of death — everlasting death — that human beings returned as earth to earth, and then all was over. Amidst thoughts morbid and impious as these were there could be nothing to console her, and she sank into the darkest depth of despair.

In these hours of deepest distress she could not weep. She thought not of the young daughters who were left to her; her husband's tears fell on her brow, but she did not look up at him; her thoughts were with her dead child; her whole heart and soul were wrapped up in recalling every reminiscence of the lost one — every syllable of his infantine prattle.

The day of the funeral came. She had not slept the night before, but towards morning she was overcome by fatigue, and sank for a short time into repose. During that time the coffin was removed into another apartment, and the cover was screwed down with as little noise as possible.

When she awoke she rose, and wished to see her child; then her husband, with tears in his eyes, told her, " We have closed the coffin — it had to be done !"

" When the Almighty is so hard on me," she exclaimed, " why should human beings be kinder?" and she burst into tears.

The coffin was carried to the grave. The inconsolable mother sat with her young daughters; she looked at them, but she did not see them; her thoughts had nothing more to do with home; she gave herself up to wretchedness, and it tossed her about as the sea tosses the ship which has lost its helmsman and its rudder. Thus passed the day of the funeral, and several days followed amidst the same uniform, heavy grief. With tearful eyes and melancholy looks her afflicted family gazed at her. She did not care for what comforted them. What could they say to change the current of her mournful thoughts?

It seemed as if sleep had fled from her for ever; it alone would be her best friend, strengthen her frame, and recall peace to her mind. Her family persuaded her to keep her bed, and she lay there as still as if buried in sleep. One night her husband had listened to her breathing, and believing from it that she had at length found repose and relief, he clasped his hands, prayed for her and for them all, then sank himself into peaceful slumber. While sleeping soundly he did not perceive that she rose, dressed herself, and softly left the room and the house, to go — whither her thoughts wandered by day and by night — to the grave that hid her child. She passed quietly through the garden, out to the fields, beyond which the road led outside of the town to the churchyard. No one saw her, and she saw no one.

It was a fine night; the stars were shining brightly, and the air was mild, although it was the 1st of September. She entered the churchyard, and went to the little grave; it looked like one great bouquet of sweet-scented flowers. She threw herself down, and bowed her head over the grave, as if she could through the solid earth

behold her little boy, whose smile she remembered so
vividly. The affectionate expression of his eyes, even
upon his sick bed, was never, never to be forgotten.
How speaking had not his glance been when she had
bent over him, and taken the little hand he was himself too weak to raise! As she had sat by his couch, so
now she sat by his grave; but here her tears might
flow freely over the sod that covered him.

"Wouldst thou descend to thy child?" said a voice
close by. It sounded so clear, so deep — its tones went
to her heart. She looked up, and near her stood a man
wrapped in a large mourning cloak, with a hood drawn
over the head; but she could see the countenance under
this. It was severe, and yet encouraging, his eyes were
bright as those of youth.

"Descend to my child!" she repeated; and there was
the agony of despair in her voice.

"Darest thou follow me?" asked the figure. "I am
Death!"

She bowed her assent. Then it seemed all at once as
if every star in the heavens above shone with the light
of the moon. She saw the many-coloured flowers on the
surface of the grave move like a fluttering garment.
She sank, and the figure threw his dark cloak round her.
It became night — the night of death. She sank deeper
than the sexton's spade could reach. The churchyard
lay like a roof above her head.

The cloak that had enveloped her glided to one side,
She stood in an immense hall, whose extremities were
lost in the distance. It was dusk around her; but before
her stood, and in one moment was clasped to her heart,

her child, who smiled on her in beauty far surpassing what he had possessed before. She uttered a cry, though it was scarcely audible, for close by, and then far away, and afterwards near again, came delightful music. Never before had such glorious, such blessed sounds reached her ear. They rang from the other side of the thick curtain — black as night — that separated the hall from the boundless space of eternity.

"My sweet mother! my own mother!" she heard her child exclaim. It was his well-known, most beloved voice. And kiss followed kiss in rapturous joy. At length the child pointed to the sable curtain.

"There is nothing so charming up yonder on earth, mother. Look, mother! — look at them all! That is felicity!"

The mother saw nothing — nothing in the direction to which the child pointed, except darkness like that of night. *She* saw with earthly eyes. She did not see as did the child whom God had called to himself. She heard, indeed, sounds — music; but she did not understand the words that were conveyed in these exquisite tones.

"I can fly now, mother," said the child. "I can fly with all the other happy children, away, even into the presence of God. I wish so much to go; but if you cry on as you are crying now, I cannot leave you, and yet I should be so glad to go. May I not? You will come back soon, will you not, dear mother?"

"Oh, stay! Oh, stay!" she cried, "only one moment more. Let me gaze on you one moment longer; let me kiss you, and hold you a moment longer in my arms."

And she kissed him, and held him fast. Then her name was called from above — the tones were those of piercing grief. What could they be?

"Hark!" said the child; "it is my father calling on you."

And again, in a few seconds, deep sobs were heard, as of children weeping.

"These are my sisters' voices," said the child. "Mother, you have surely not forgotten them?"

Then she remembered those who were left behind. A deep feeling of anxiety pervaded her mind; she gazed intently before her, and spectres seemed to hover around her; she fancied that she knew some of them; they floated through the Hall of Death, on towards the dark curtain, and there they vanished. Would her husband, her daughters, appear there? No; their lamentations were still to be heard from above. She had nearly forgotten them for the dead.

"Mother, the bells of heaven are ringing," said the child. "Now the sun is about to rise."

And an overwhelming, blinding light streamed around her. The child was gone, and she felt herself lifted up. She raised her head, and saw that she was lying in the churchyard, upon the grave of her child. But in her dream God had become a prop for her feet, and a light to her mind. She threw herself on her knees and prayed: —

"Forgive me, O Lord my God, that I wished to detain an everlasting soul from its flight into eternity, and that I forgot my duties to the living Thou hast graciously spared to me!"

And as she uttered this prayer it appeared as if her heart felt lightened of the burden that had crushed it. Then the sun broke forth in all its splendour, a little bird sang over her head, and all the church bells around began to ring the matin chimes. All seemed holy around her; her heart seemed to have drunk in faith and holiness; she acknowledged the might and the mercy of God; she remembered her duties, and felt a longing to regain her home. She hurried thither, and leaning over her still sleeping husband, she awoke him with the touch of her warm lips on his cheek. Her words were those of love and consolation, and in a tone of mild resignation she exclaimed, —

"God's will is always the best!"

Her husband and her daughters were astonished at the change in her, and her husband asked her, —

"Where did you so suddenly acquire this strength — this pious resignation?"

And she smiled on him and her daughters as she replied, —

"I derived it from God, by the grave of my child."

Q

Charming.

THE sculptor Alfred — surely you know him? We all know him. He used to engrave gold medallions; went to Italy and returned again. He was young then; indeed, he is young now, though about half a score of years older than he was at that time.

He returned home, and went on a visit to one of the small towns in Zealand. The whole community knew of the arrival of the stranger, and who he was. There was a party given on his account by one of the richest families in the place; every one who was anybody, or had anything, was invited; it was quite an event, and the whole town heard of it without beat of drum. A good many apprentice boys and poor people's children, with a few of their parents, ranged themselves outside, and looked at the windows with their drawn blinds, through which a blaze of light was streaming. The watchman might have fancied he had a party himself, so many people occupied his quarters in the street. They all seemed merry on the outside; and in the inside of the house everything was pleasant, for Herr Alfred, the sculptor, was there.

He talked, and he told anecdotes, and every one present listened to him with pleasure and deep attention, but no one with more eagerness than an elderly widow of good standing in society; and she was, in reference to all that Herr Alfred said, like a blank sheet of whity-brown paper, that quickly sucks the sweet things in, and is ready for more. She was very susceptible, and totally ignorant — quite a female Caspar Hauser.

"I should like to see Rome," said she. "That must be a charming town, with the numerous strangers that go there. Describe Rome to us now.· How does it look as you enter the gate?"

"It is not easy to describe Rome," said the young sculptor. "It is a very large place; in the centre of it stands an obelisk, which is four thousand years old."

"An organist!" exclaimed the astonished lady, who had never before heard the word *obelisk*.

Many of the party could scarce refrain from laughing, and among the rest the sculptor. But the satirical smile that was gathering round his mouth glided into one of pleasure; for he saw, close to the lady, a pair of large eyes, blue as the sea. They appertained to the daughter of the talkative dame, and when one has such a daughter one could not be altogether ridiculous. The mother was like a bubbling fountain of questions, constantly pouring forth; the daughter, like the fountain's beautiful naiad, listening to its murmurs. How lovely she was! She was something worth a sculptor's while to gaze at; but not to converse with; and she said nothing, at least very little.

"Has the Pope a great family?" asked the widow.

And the young man answered as if the question might
have been better worded, —

"No, he is not of a high family."

"I don't mean that," said the lady; "I mean has he a
wife and children?"

"The Pope dare not marry," he replied.

"I don't approve of that," said the lady.

She could scarcely have spoken more foolishly, or
asked sillier questions; but what did all that signify when
her daughter looked over her shoulder with that most
winning smile?

Herr Alfred talked of the brilliant skies of Italy, and
its cloud-capped hills; the blue Mediterranean; the soft
South; the beauty which could only be rivalled by the
blue eyes of the females of the North   And this was
said pointedly; but she who ought to have understood it
did not allow it to be seen that she had detected any compliment in his words, and this was also charming.

"Italy!" sighed some.   "Travelling!" sighed others.
"Charming, charming!"

"Well, when I win the fifty-thousand-dollar prize in
the lottery," said the widow, "we shall set off on our
travels too — my daughter and I; and you, Herr Alfred,
shall be our escort. We shall all three go, and a few
other friends will go with us, I hope;" and she bowed
invitingly to them all round, so that each individual might
have thought, "It is I she wishes to accompany her."
"Yes, we will go to Italy, but not where the robbers are;
we will stay in Rome, or only go by the great high roads,
where people are safe, of course."

And the daughter heaved a gentle sigh.   How much

can there not lie in a slight sigh, or be supposed to lie in it! The young man put a world of feeling into it; the two blue eyes that had beamed on him that evening concealed the treasure — the treasure of heart and of mind, richer far than all the glories of Rome; and when he left the party he was over head and ears in love with the widow's pretty daughter.

The widow's house became the house of all others most visited by Herr Alfred, the sculptor. People knew that it could not be for the mother's sake he sought it so often, although he and she were always the speakers; it must be for the daughter's sake he went. She was called Kala, though christened Karen Malene: the two names had been mutilated, and thrown together into the one appellation, *Kala.* She was very beautiful, but rather silly, some people hinted, and rather indolent. She was certainly a very late riser in the morning.

" She has been accustomed to that from her childhood," said her mother. " She has always been such a little Venus that she was scarcely ever found fault with. She is not a very early riser, but to this she owes her fine clear eyes."

What power there was in these clear eyes — these swimming blue eyes! The young man felt it. He told anecdote upon anecdote, and answered question after question; and mamma always asked the same lively, sensible, pertinent questions as she had asked at first.

It was a pleasure to hear Herr Alfred speak. He described Naples, the ascent of Mount Vesuvius, and several of its eruptions; and the widow lady, who had never heard of them before, was lost in surprise.

"Mercy on us!" she exclaimed; "then it is a volcano? Does it ever do any harm to anybody?"

"It has destroyed entire towns," he replied: "Pompeii and Herculaneum."

"But the poor inhabitants! Did you see it yourself?"

"No, not either of these eruptions, but I have a sketch taken by myself of an eruption which I did witness."

Then he selected from his portfolio a sketch done with a black-lead pencil; but mamma, who delighted in highly-coloured pictures, looked at the pale sketch, and exclaimed in amazement,—

"You saw it gush out white?"

Mamma got into Herr Alfred's black books for a few minutes, and he felt profound contempt for her; but the light from Kala's eyes soon dispelled his gloom. He bethought him that her mother had no knowledge of drawing, that was all; but she had what was far better — she had the sweet, beautiful Kala.

As might have been expected, Alfred and Kala became engaged, and their betrothal was announced in the newspaper of the town. Mamma bought thirty copies of it, that she might cut the paragraphs out, and inclose them to various friends. The betrothed pair were very happy, and so was the mamma: she felt almost as proud as if her family were going to be connected with Thorwaldsen.

"You are his successor at any rate," she said; and Alfred thought that she had said something very clever. Kala said nothing, but her eyes brightened, and a lovely smile played around her well-formed mouth. Every movement of hers was graceful: she was very beautiful — that cannot be said too often.

Alfred was making busts of Kala and her mother; they sat for him, and saw how with his finger he smoothed and moulded the soft clay.

"It is a compliment to us," said his mother-in-law elect, "that you condescend to do that simple work yourself, instead of letting your men dab all that for you." ·

"No; it is absolutely necessary that I should do this myself in the clay," he replied.

"Oh! You are always so exceedingly gallant!" said mamma; and Kala gently pressed his hand, to which pieces of clay were sticking.

He discoursed to them about the magnificence of Nature in its creations, the superiority of the living over the dead, plants over minerals, animals over plants, human beings over mere animals; how mind and beauty manifested themselves through form, and that the sculptor sought to bestow on his forms of clay the greatest possible beauty and expression.

Kala remained silent, revolving his words. Her mother said, "It is difficult to follow you; but though my thoughts go slowly, I hold fast what I hear."

And the power of beauty held him fast; it had subdued him — entranced and enslaved him. Kala's beauty certainly was extraordinary; it was enthroned in every feature of her face, in her whole figure, even to the points of her fingers. The sculptor was bewildered by it; he thought only of her — spoke only of her; and his fancy endowed her with all perfection.

Then came the wedding-day, with the bridal gifts and the bride's-maids; and the marriage ceremony was duly performed. His mother-in-law had placed in the room where the bridal party assembled the bust of Thorwald-sen, enveloped in a dressing-gown. " He ought to be a guest, according to her idea," she said. Songs were sung, and healths were drunk. It was a handsome wedding, and they were a handsome couple. " Pygmalion got his Galathea," was a line in one of the songs.

"That was something from mythology," remarked the widow.

The following day the young couple started for Copenhagen, where they intended to reside ; and the mamma accompanied them, to give them a helping hand, she said, which meant to take charge of the house. Kala was to be a mere doll. Everything was new, bright, and charming. There they settled themselves all three; and Alfred, what can be said of him, only that he was like a bishop among a flock of geese?

The magic of beauty had infatuated him. He had gazed upon the case, and not thought of what was in it; and it is unfortunate, very unfortunate, in the marriage state. When the case decays, and the gilding rubs off, one then begins to repent of one's bargain. It was very mortifying to Alfred that in society neither his wife nor his mother-in-law was capable of entering into general conversation — that they said very silly things, which, with all his wittiest efforts, he could not cover.

How often the young couple sat hand in hand, and he spoke, and she dropped a word now and then, always in the same tone, like a clock striking one, two, three! It was quite a relief when Sophie, a female friend, came.

Sophie was not very pretty; she was slightly awry, Kala said; but this was not perceptible except to her female friends. Kala allowed that she was clever. It never occurred to her that her talents might make her dangerous. She came like fresh air into a close, confined puppet show; and fresh air is always pleasant. After a time the young couple and the mother-in-law went to breathe the soft air of Italy. Their wishes were fulfilled.

"Thank Heaven, we are at home again!" exclaimed both the mother and the daughter, when, the following year, they and Alfred returned to Denmark.

"There is no pleasure in travelling," said the mamma; "on the contrary, it is very fatiguing — excuse my saying so. I was excessively tired, notwithstanding that I had my children with me. And travelling is extremely expensive. What hosts of galleries you have to see! What quantities of things to be rushing after! And you are so teased with questions when you come home, as if it were possible to know everything. And then to hear that you have just forgotten to see what was most charming! I am sure I was quite tired of these everlasting Madonnas; one was almost turned into a Madonna one's self."

"And the living was so bad," said Kala.

"Not a single spoonful of honest meat soup," rejoined the mamma. "They dress the victuals so absurdly."

Kala was much fatigued after her journey. She continued very languid, and did not seem to rally — that was the worst of it. Sophie came to stay with them, and she was extremely useful.

The mother-in-law allowed that Sophie understood household affairs well, and had many accomplishments,

12

which she, with her fortune, had no need to trouble herself about; and she confessed, also, that Sophie was very estimable and kind. She could not help seeing this when Kala was lying ill, without making the slightest exertion in any way.

If there be nothing but the case or framework, when it gives way it is all over with the case. And the case had given way. Kala died.

" She was charming!" said her mother. " She was very different from all these antiquities that are half mutilated. Kala was a perfect beauty!"

Alfred wept, and his mother-in-law wept, and they both went into mourning. The mamma went into the deepest mourning, and she wore her mourning longest. She also retained her sorrow the longest; in fact, she remained weighed down with grief until Alfred married again. He took Sophie, who had nothing to boast of in respect to outward charms.

"He has gone to the other extremity," said his mother-in-law; "passed from the most beautiful to the ugliest. He has found it possible to forget his first wife. There is no constancy in man. My husband, indeed, was different; but he died before me."

" Pygmalion got his Galathea," said Alfred. " These words were in the bridal song. I certainly did fall in love with the beautiful statue that became imbued with life in my arms. But the kindred soul, which Heaven sends us, one of those angels who can feel with us, think with us, raise us when we are sinking, I have now found and won. You have come, Sophie, not as a beautiful form, fascinating the eye, but prettier, more pleasing than was necessary. You excel in the main point. You have come and taught the sculptor that his work is but clay — dust; only a copy of the outer shell of the kernel we ought to seek. Poor Kala! her earthly life was but like a short journey. Yonder above, where those who sympathise shall be gathered together, she and I will probably be almost strangers."

"That is not a kind speech," said Sophie; "it is not a Christian one. Up yonder, where 'they neither marry nor are given in marriage,' but, as you say, where spirits shall meet in sympathy — there, where all that is beautiful shall unfold and improve, her soul may perhaps appear so glorious in its excellence that it may far outshine mine and yours. You may then again exclaim, as you did in the first excitement of your earthly admiration, 'Charming — charming!'"

THE END.

Cambridge : Electrotyped and Printed by Welch, Bigelow, & Co.